12

Horological Virtuosos

Elizabeth Doerr
Ralf Baumgarten

teNeues

Elizabeth Doerr

Born on June 14, 1966, Elizabeth Doerr has dedicated most of her waking hours to mechanical watches since 1991. She is a regular contributor to *International Watch, DuPont Registry, Cigar Aficionado, Elite Traveler,* and other U.S. publications as well as internationally based magazines such as *Plaza Watch,* Mexico's *Tiempo de Relojes,* Holland's *Watch,* and Swiss website worldtempus.com. Doerr is also the editor of *Wristwatch Annual,* a yearly 400-page overview of high-end mechanical watches.

This native of Petoskey, Michigan's independent expertise and standing in her field are unique: she is a member of several international juries as well as the Time Aeon Foundation's Pleiad Competition, initiated by Harry Winston. Having co-authored *Elite Timepieces* in 2008, Doerr is also currently working on further book projects.

Doerr moved to Germany in 1988 to pursue graduate studies in German language and literature after graduating from Kalamazoo College with a bachelor's degree in German and European Studies.

She now lives in Karlsruhe, on the northern edge of Germany's Black Forest, with her husband and two children.

Ralf Baumgarten

Ralf Baumgarten was born on September 15, 1960, in Saarbrücken, Germany. He studied communication design at the University of Wuppertal, earning his diploma in 1988. In 1984, he was awarded top prizes in Polaroid and Kodak competitions.

Baumgarten is personally interested in watchmaking, and his work boasts a varied repertoire of themes seen in his exhibitions: "Stuhl in 30 Akten," Cologne, 1997; the Volkenroda convent in Weimar as the subject of an exposition of the 2000 World's Fair; "Au Coeur du Val," an exhibition focusing on a Swiss valley that took place in Hérémence in 2007; "See the Flames," shown in Sulzbach, Germany, in 2008; and faces of camels in Dubai for a widely publicized desert-themed project that was exhibited in 2009.

Baumgarten's first book on the subject of watchmaking, *UhrMenschen,* received a Red Dot design award in 2005. His artful photography has been featured in important publications all over the world such as *GEO, Manager Magazin, VIEW,* and *Tiempo de Relojes.* Baumgarten has also photographed for a number of reputable companies, among them DuPont, Fujitsu Siemens, and WMF.

Baumgarten now lives in Cologne, Germany, with his wife and two children.

12

Twelve Faces of Time
Horological Virtuosos

Elizabeth Doerr
Ralf Baumgarten

teNeues

Contents

Vyskocil

Preface

A wristwatch is by no means simply an interchangeable object. And even though they fall under the same moniker, a handmade watch should really be considered a separate entity from a mass-produced timepiece.

For passionate watchmakers and collectors, any watch must be meaningful and tell a story of its own. Timekeeping for the sake of timekeeping can be found these days in one's mobile phone, automobile dashboard, or even microwave oven. The watch must therefore be a cultural item, a symbol of one's nature, an expression of timelessness that happens to keep accurate time.

Time passes, time is captured, time flies–and in no other place does it do those things better than in the hands of the twelve virtuosos whose narratives make up the substance of this book. Each of them can well be considered an epoch-making watchmaker in his own right, and they would probably have been so titled had they lived a century earlier. They all have their very own story to tell: in the hands of these masters, time becomes tangible, perpetual, and personal–art as opposed to contingency.

And nowhere is this better illustrated than in this volume, which combines unusual horological stories, that watchmaking is universal: it knows no national boundaries and can really no longer be considered purely Swiss, even if most of today's meaningful watchmaking does take place in Switzerland. These twelve watchmakers represent seven different nations, though nine of their places of business are now mainly concentrated on the 41,000 square meters of that trilingual Helvetian land officially known as the Swiss Confederation. Language is not a barrier; the language of watchmaking is universal.

Nine of the watchmakers whose stories we tell here are or were members of the A.H.C.I. (Académie Horlogère des Créateurs Indépendents, or the Horological Academy of Independent Creators), a loose grouping of independent watchmakers founded in 1985 by Vincent Calabrese and Svend Andersen who have decided to forego the usual route of working for a luxury brand in favor of creating their own art. The reasons for this vary, but the horological world is definitively the richer for it; many of the timepieces these masters are famous for would not have seen the light of day had they been conceived in the environs of the corporate world.

The two watchmakers featured in this book who are geographically the farthest from Switzerland happen to be two who do not belong to the A.H.C.I.: Kenji Shiohara and Roger Smith. Shiohara, a product of a corporate environment and a strict Japanese culture, did not have the same luxury of independence available to the others; his history is all the more incredible for it. Smith, rather isolated on the Isle of Man, simply does not see the need. Rooted more in traditional English watchmaking than Swiss, Smith has essentially come into his own by relying upon himself rather than the prevalent Swiss system.

The other ten watchmakers–even though Dr. Ludwig Oechslin is also not a member of the A.H.C.I.–enjoy more of a give-and-take with the established system in many ways.

The A.H.C.I. has been a wonderful tool for them to draw more attention to their artisan styles of watchmaking and attract broader audiences than ever before. With the twentieth anniversary of the creative group having just passed, and *haute horlogerie* once again in more corporate spotlights, it is natural that this niche of the watchmaking industry should attract more attention than ever.

The past and future are two elements that might seem separate in watchmaking, but couldn't be more related. They need to be in harmony for modern watchmaking to maintain its rightful place in history and the arts.

For five centuries, the measurement of time has enchanted mankind and fascinated the soul. The modern age of the post-mechanical Renaissance couldn't be a more fortuitous one for a creative watchmaker to live in. While past greats like Antide Janvier, Thomas Mudge, Ferdinand Berthoud, Jost Bürgi, and, of course, Abraham-Louis Breguet have influenced modern watchmaking technology, just think what they might have been able to accomplish had they had the benefit of Einstein's theories and computer-assisted precision. Though they do benefit from such modern advantages, the twelve watchmakers presented here are characterized by the genius that naturally emanates from them. Having studied the masters, they know how to arrive at their own conclusions. Having the benefit of technology in the modern age, not to mention modern media, they have the opportunity to influence generations of passionate fans of *haute horlogerie* and make a living from their art.

These twelve faces of time are arranged in descending order of age, symbolizing the double-headed view of a horological Janus: the past of watchmaking inspiring the future. Therefore, *12 Faces of Time* starts out with Philippe Dufour, a watchmaker who humbly feels the need to teach upcoming generations and pass on the great cache of knowledge he has amassed over the course of a life dedicated to ticking and chiming.

Dr. Ludwig Oechslin and Paul Gerber are tied to each other with more than just the similarities of proximity in age and an all-encompassing curiosity. Their mutual respect and passion for the subject has led them to work together on various projects such as the MIH Watch and the recently released Ochs und Junior.

François-Paul Journe's success in the world of watchmaking is more than just an inspiration to his fellow independents: his rise to fame based largely on his tremendous talent has directed collectors' avid attention to independent watchmaking as a whole. Journe's company, which employs about sixty technicians completing something like 750 timepieces each year, is a prime example of the endless possibilities that independently minded watchmakers with a clever head for business and unparalleled talent can achieve.

One of Dufour's most interested and interesting "protégés" follows Journe in our order of things: Kenji Shiohara initiated a small division for high watchmaking within the corporate culture at Seiko, creating the best of both traditional and highly mechanized worlds. Dufour is most widely known for his meticulous attention to finishing, which he completes strictly manually utilizing only handheld tools that have been in use for hundreds of years. Shiohara's creations, on the other hand, couldn't be further from the norm and literally merge electronics with tradition. Had the idea occurred to them in the 1960s, the fathers of quartz-controlled timekeeping would certainly have immediately understood the inspiration behind the creations that emerge from Shiohara's Micro Artist Studio.

Kari Voutilainen's immense courage saw him stepping out from behind the shadow of a successful decade instructing at WOSTEP and joining the workforce at Parmigiani Fleurier. A man whose vocation is truly his passion, Voutilainen knew that the only way to quiet the ferment of ideas for creative timepieces bubbling inside of him–and to stop himself from working uninterruptedly–was to establish his own company. A favorite among collectors and colleagues for his easygoing manner, humble attitude, and technically beautiful wristwatches, Voutilainen enriches the Swiss watchmaking landscape so reminiscent of his homeland.

Vianney Halter is perhaps the forerunner of the postmodern era of watchmaking. It is Halter who gave expression to unusual forms within the confines of traditional watchmaking and set off a trend for other freethinkers such as Maximilian Büsser with his MB&F and, intuitively speaking, even companies such as Urwerk, which in turn has inspired other progressive mini-brands in this anachronistic modern era. Luckily for the watch collector, Halter's passion knows no limits, and it is impossible to say what will emerge from its depths next.

Beat Haldimann keeps a low profile, but this cloud has a silver lining for collectors of his timepieces–which include interesting resonance-driven wall clocks–as they generally must make a trip to his Thun-based workshop to see him. And this trip is well worth the effort: the visitor is afforded a serious look into Haldimann's seemingly idyllic daily routine against the backdrop of Thun's breathtaking scenery and the particularly beautiful workshop-cum-abode only overshadowed by the timepieces themselves.

Volker Vyskocil is the one watchmaker in this book who does not have a traditional watchmaker's education: after earning a diploma in mechanical engineering, he decided to switch gears and taught himself all he would need to know about watchmaking from books. The look of his work reflects this fact as it mirrors his passion for technical subjects. His many speaking engagements within the watch world confirm the respect he has earned from colleagues and collectors alike. An idealist refusing to compromise, Vyskocil really lives his art and is an inspiration to all who come in contact with him.

The flying tourbillons of Thomas Prescher are perhaps only trumped by the gracefully moving arms of the *bras en l'air* figures displaying the time on his Tempusvivendi models. Tempusvivendi, "living time," is precisely what Prescher does in his picturesque Twann workshop overlooking Lake Biel, a place where time seems to stand still, though he never does unless seated at his workbench.

12 Faces of Time finishes up with Roger Smith and Felix Baumgartner, who could well be considered the faces of watchmaking's future, the former representing a more historical tradition than almost any other living, working watchmaker and the latter standing for the myriad possibilities that the future may bring. Jointly they easily embody the animated yin-yang that is modern mechanical watchmaking.

To close out this book, we have compiled a few short facts about each of these watchmakers and list their works in a "watchography," similar to what one finds in reference books on musicians. We also share some of the awards these twelve have amassed, though prizes given by magazines have been omitted: Gerber, Oechslin, Dufour, Haldimann, and Journe have received the Prix Gaïa for excellence in watchmaking; Voutilainen, Halter, and Journe have won various honors at watchmaking's Oscars, the Grand Prix d'Horlogerie de Genève. Additionally, three of these artists were chosen by Harry Winston Rare Timepieces' then-CEO Maximilian Büsser to create one of the celebrated Opus models, a phenomenal outlet to a broader audience that edged independent watchmaking even further into the spotlight: Journe (Opus 1), Halter (Opus 3), and Baumgartner (Opus 5).

Once considered scientists–as were many of the role models of bygone days of our twelve virtuosos–today a watchmaker is someone who basically changes batteries on quartz timepieces. Proudly, these twelve belong to an art that almost died out thirty years ago and now enjoys only minimal celebrity. These twelve visages of timekeeping continue to reach into the magic hat of scientific horology to create some of the most unique and complicated watchmaking available in the modern age.

Both Janvier and Breguet researched the phenomenon of resonance: their work is carried on today by the findings and creations of Journe, Haldimann, and Dufour. Resonance in watchmaking occurs when one moving object's vibrations are picked up by another object, which begins to vibrate at the same speed. Horologically, this takes place when two balance wheels placed near each other alternately act as exciter and resonator. Among the people involved in the most extreme *haute horlogerie* today, we find that each one of our twelve watchmakers acts as an exciter in his own way–initiating resonance in many corners of the industry, whether they were personally involved or not. By harmonizing, they improve both accuracy and creativity.

And let's not forget that Breguet patented the tourbillon in 1801; this too–until recently the highest expression of a watchmaker's skill–is a theme that has been expounded upon by many of the twelve here: Journe first put a constant force tourbillon into a wristwatch in 1991; Prescher completed a full trilogy of single, double, and triple axis tourbillons to make history; Oechslin turned a karussel into the hour hand of a watch and added a new escapement form to it; Haldimann centralized his floating, flying, humming tourbillon and transformed it into a kinetic sculpture; Gerber added a flying tourbillon to a century-old movement; Voutilainen proved his immense talent with a tourbillon pocket watch in 1994; and Smith's first handmade timepiece–which contained this rotating escapement–impressed his legendary mentor so much so that he was allowed to join his innermost circle.

Tenacity, creativity, courage: these artisans all possess it. Watchmaking is a philosophy unto itself, but without these characteristics our twelve protagonists would certainly have never had the chance to name such unique products after themselves–and enrich the world of horology.

The journeys Ralf Baumgarten and I have made to discover new faces and deepen our understanding of familiar ones took us not only to various exotic locations in a total of four countries, but on a journey inside ourselves to understand what this even dozen of unique horologists have collectively created with their individual styles of mechanical art. The journey has seen both of us grow, understanding not only more about ourselves and our chosen professions, but what it really is that makes us and each one of these tremendous watchmakers tick.

We could not be more honored to call many of these exquisite watchmakers our friends and to honor their creativity and timeless work with this publication. We only hope that you have as much enjoyment in reading these words and viewing these photographs as we did in gathering the information and creating images of brief seconds we hope represent the true meaning of these men's work. We know that the timepieces they create will be handed down for generations as treasured heirlooms, the owners of which most certainly also share the same sort of unique relationship to these pioneers as we do. Hats off to their courage to be different and desire to create what is inside of them–so inspiring to generations of apprentices and watch aficionados of the past, present, and most certainly the future.

Elizabeth Doerr
Ralf Baumgarten
2008–2009

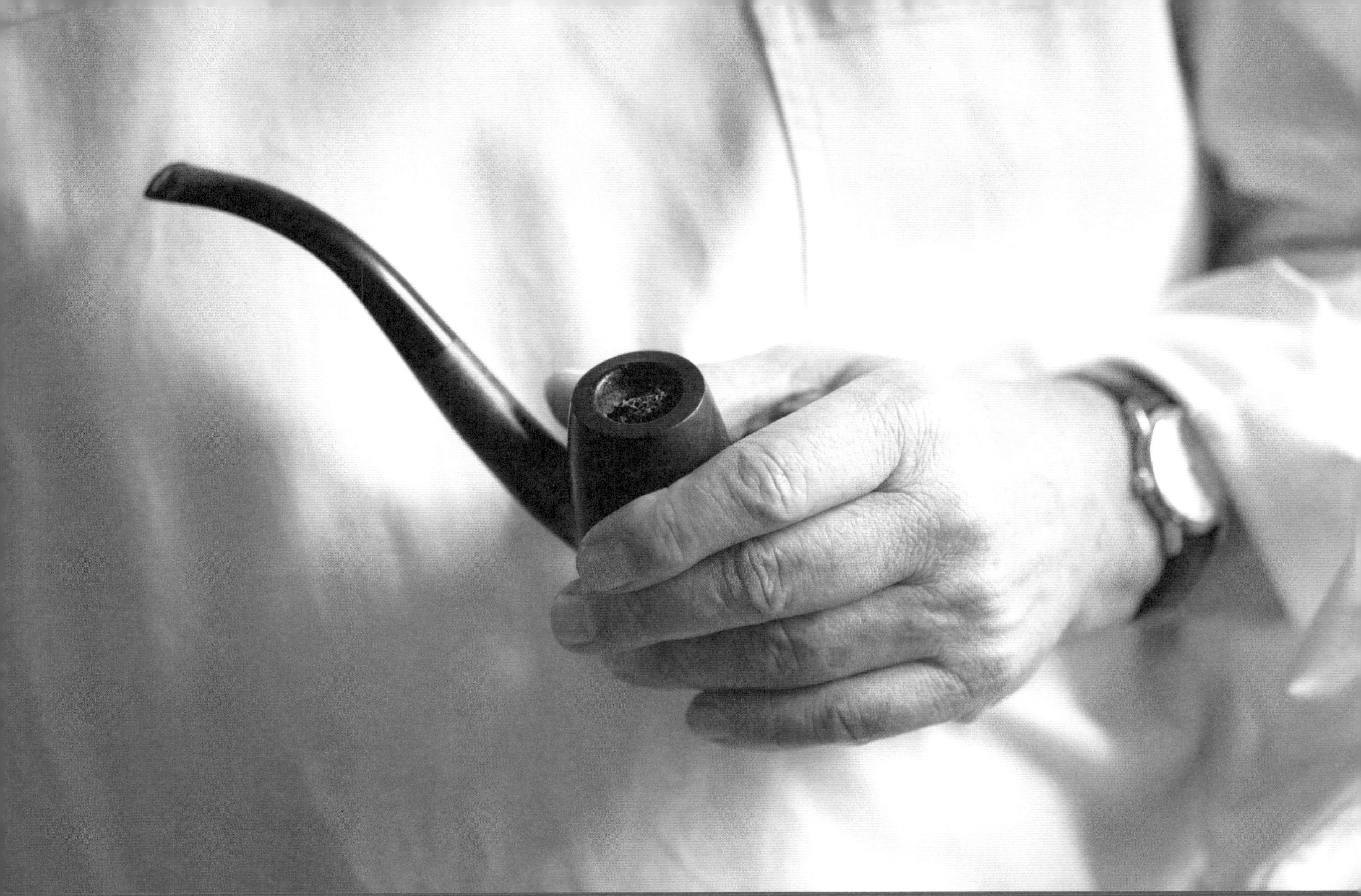

Philippe Dufour

No one finishes a movement the way Philippe Dufour does. Anachronistic, perhaps too much a perfectionist, he spends his days cutting, filing, grinding, and polishing until everything is just right. There is not another watchmaker with Dufour's ultimate patience. This patience was born of necessity, for this watchmaker had to follow a rather long and winding path to reach his current status as a living legend in horology.

The situation in the 1980s was an unpleasant one for Swiss watchmaking in general, and Dufour struggled to make ends meet. An artist and a creator, marketing was not his forte. A little bird urged him to approach a large and reputable brand about buying a minute repeater he had already spent 2,000 hours on. The brand accepted his proposal with open arms, and Dufour set to work on five pieces for the new contractor. He was sorely disappointed, though, when he discovered that he would not be receiving credit for these masterpieces, the brand adapting the usual attitude of the time of attempting to convince the consumer that the company did everything itself.

"This was tough, you know?" Dufour explains now. "But it helped me to start making watches, it helped me to survive during these five years. They didn't respect my work–and when you spend one year on a watch, you expect respect. This one watch nearly destroyed me."

At this same time, Dufour–who learned watchmaking in the traditional way as it has been practiced for over 400 years–saw the necessity of acquiring the skills to design his movements on the computer, a commonplace tool today. After two and a half years spent learning to use AutoCAD, Dufour was finally able to present his own watch at the Basel Fair in 1992, a grande et petite sonnerie minute repeater wristwatch.

Today, Dufour has a practical attitude toward new technologies. "I would say I don't reject any technology. I try to find what I think would be the best technology for what I am doing. But, you know, the value of the product is in fact what we add [with our own hands]. If you have a component, for example, coming from a CNC machine, it's what you do to that component, you know? And, in fact, the value of the product is–beside the cost of the platinum case, say–what we add, in time and to make things perfect: the beveling, the edges, the sinks around the jewels, everything. We spend hours and hours doing that. What is pleasant is that people are recognizing what we've done and what is different about it."

Dufour feels himself to be a distinct part of the endless chain of watchmakers that make up the body of horological history in its totality. Each generation of watchmakers brings something new to the mix, but Dufour, perhaps more than any other master in the field, champions the importance of taking a step back, returning to those original values of Swiss watchmaking that have made it the grand institution it is today.

PHILIPPE DUFOUR
Horlogerie compliquée

His contribution to this continuity does not have patents attached to it–Dufour does not patent anything, as he firmly believes that there is almost never anything new in watchmaking. The fine quality of his meticulous work is second to none in this industry, and it is this bar he has unwittingly set that represents his greatest legacy; Dufour's work is the target that watchmakers in search of perfection aspire to reach.

PHILIPPE DUFOUR
METALEM
SWISS MADE

I would say my goal is probably to try to slow down a certain process, the process of what we are facing today," Dufour explains his style of watchmaking and the general loss of handcrafted arts at the same time. "When I started to work alone, the biggest problem I was facing was *how* to do things. There is nothing written. So you have to try, to test, everything–because it has not been transmitted. There is more theory and less practice in professions now, you know? We are losing that process. And I try, in my own way, to stop it–no, not stop it, but to slow it down."

Above and beyond the traditional way Dufour practices his craft, his main goal remains the same: "It's to stay humble and try to do it better every day. To improve every day." Watchmaking is life to Dufour: a watch is a living thing that he breathes vitality into. "Besides showing the time, you have something with life. And people are now rediscovering that. The fact of winding a watch–take [my] Simplicity, start to wind it, and...it's magic. It's a relationship between the owner and the watch."

Dufour also has a singular relationship to his timepieces. "Especially when I start a new project, and I make my first watch. The first time I had the Duality running with its two balances, it was magic. You bring motion, you bring life."

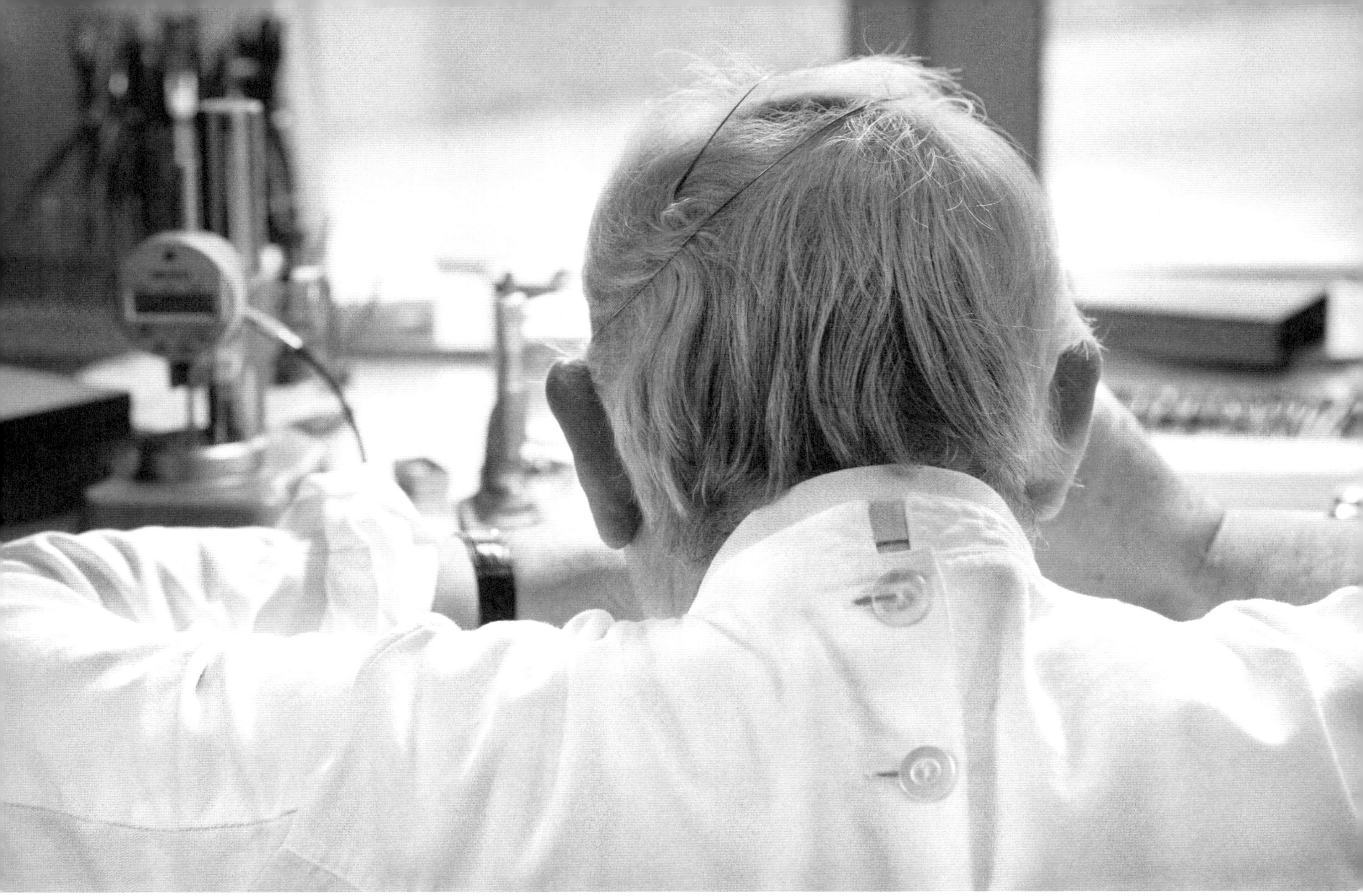

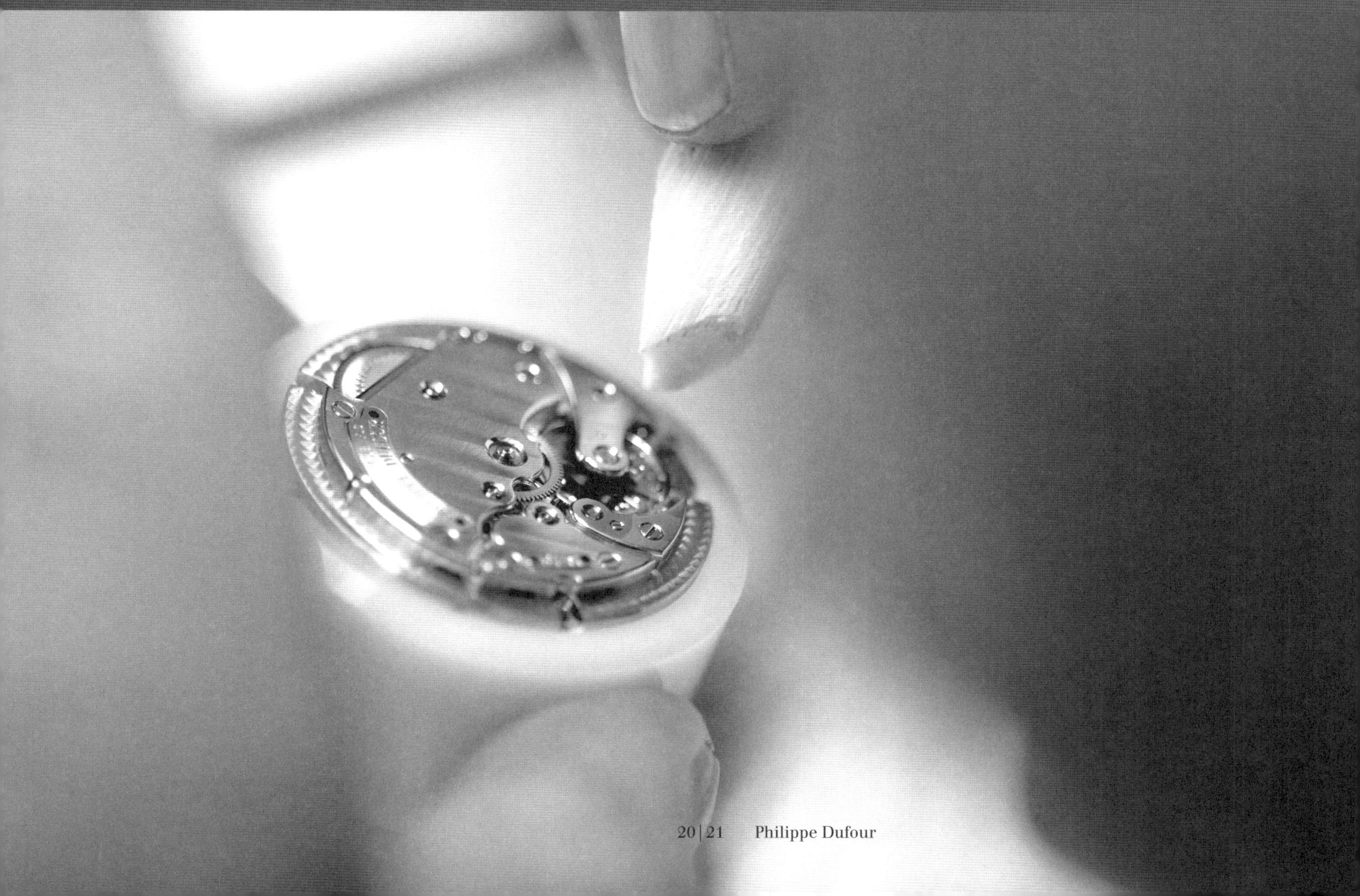

"Somebody decided that to play music we need some notes; these notes are universal today. And with those notes you can play any kind of music – jazz, classical, rap – any kind of music. It's all with the same notes. Watchmaking is the same thing, but we don't call them notes. Harrison, Breguet… it's all these people. They found all the notes and made real inventions, and today what we do is play those notes."

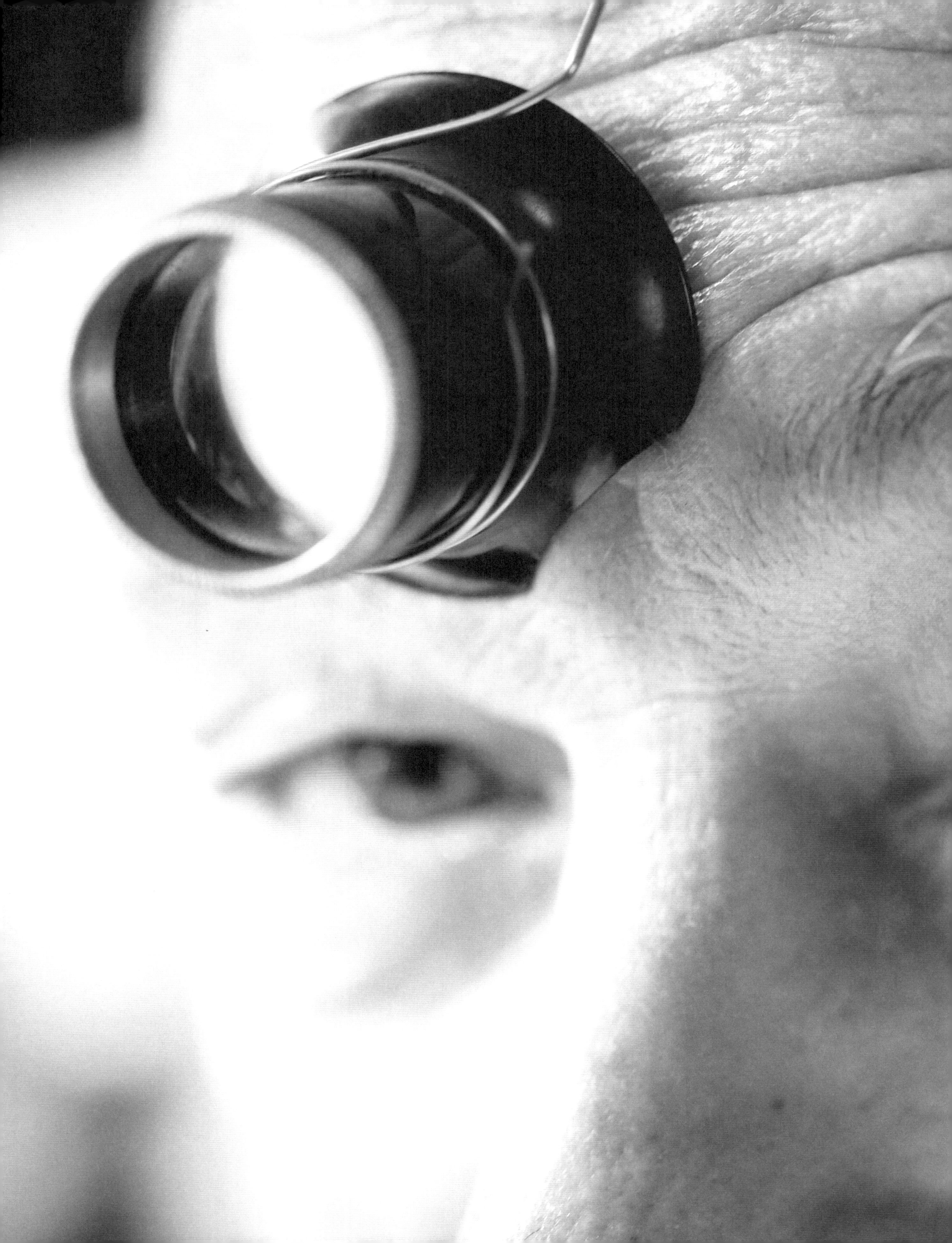

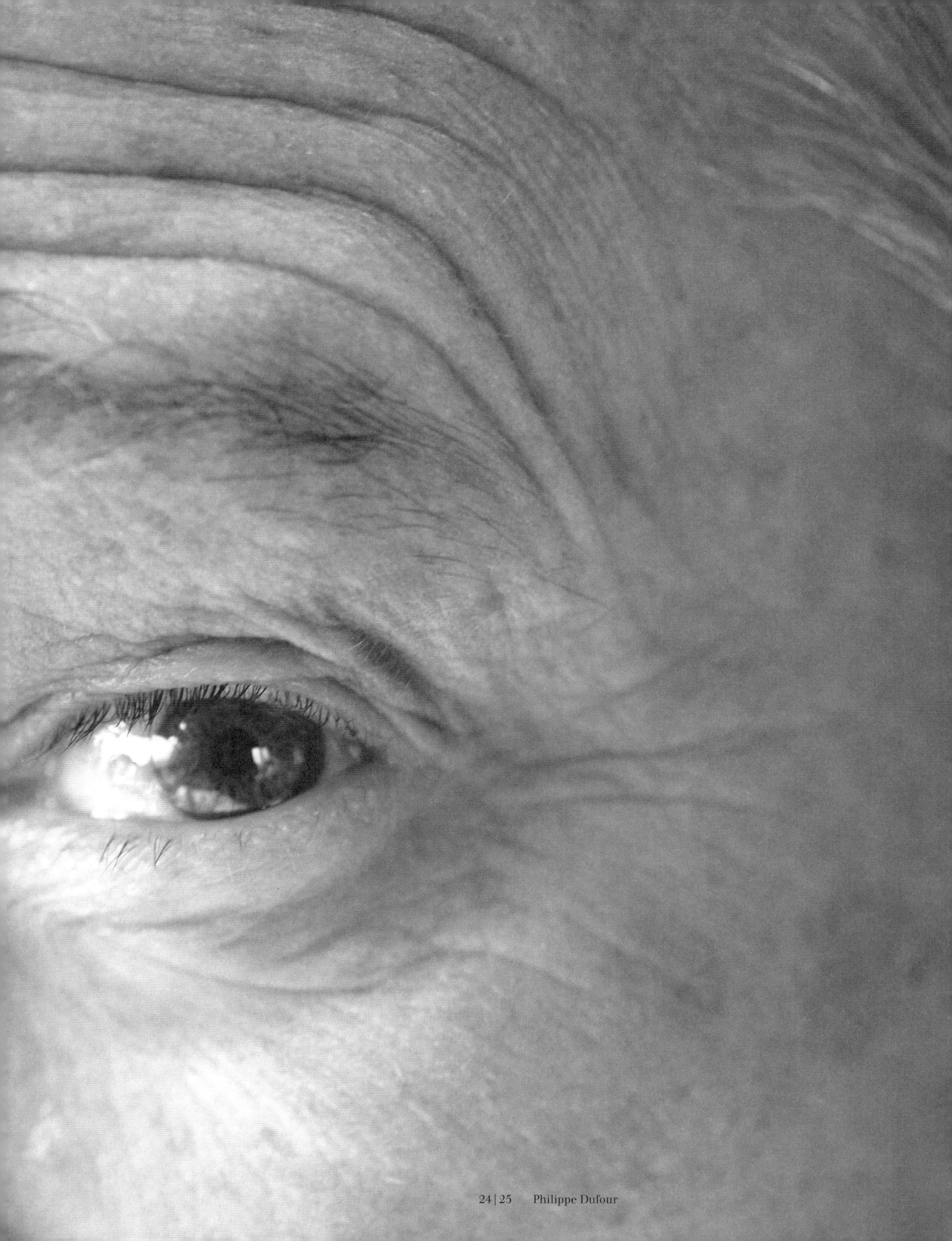

Nothing is impossible: this is the base element of Paul Gerber's concept. If Gerber had not embraced this attitude right from the beginning, he would not be half the watchmaker he is today. Being known for his ability to achieve the seemingly unachievable is not just a reputation, but a true facet of Gerber's entire horological being.

For Gerber, the word *impossible* does not exist. If he has an idea, or if someone else comes to him with an idea they would like him to realize, he does just that. "That is my goal, what I try to achieve, and for me it is normal," he concedes in his modest way. And this can take on many faces: altering a 150-year-old movement that Franck Muller had claimed as his own at the request of a collector and turning it into a creative collaboration of three acclaimed watchmakers over a 150-year span or turning a tourbillon movement into a unique Russian work of art the likes of which has never been seen before, Gerber is the man for these jobs–just to see if he can.

Paul Gerber

ZH·174723

In fact, most of the things he does are such a testing of boundaries. It's not about more precision or better reliability, though these are elements that need to be present. For Gerber it is indeed just to explore whether he can do it. It is not Gerber's humble way to doubt what others have done, but rather to prove to himself if he can realize them in a different way–not necessarily better, and he does underscore this difference, but simply different.

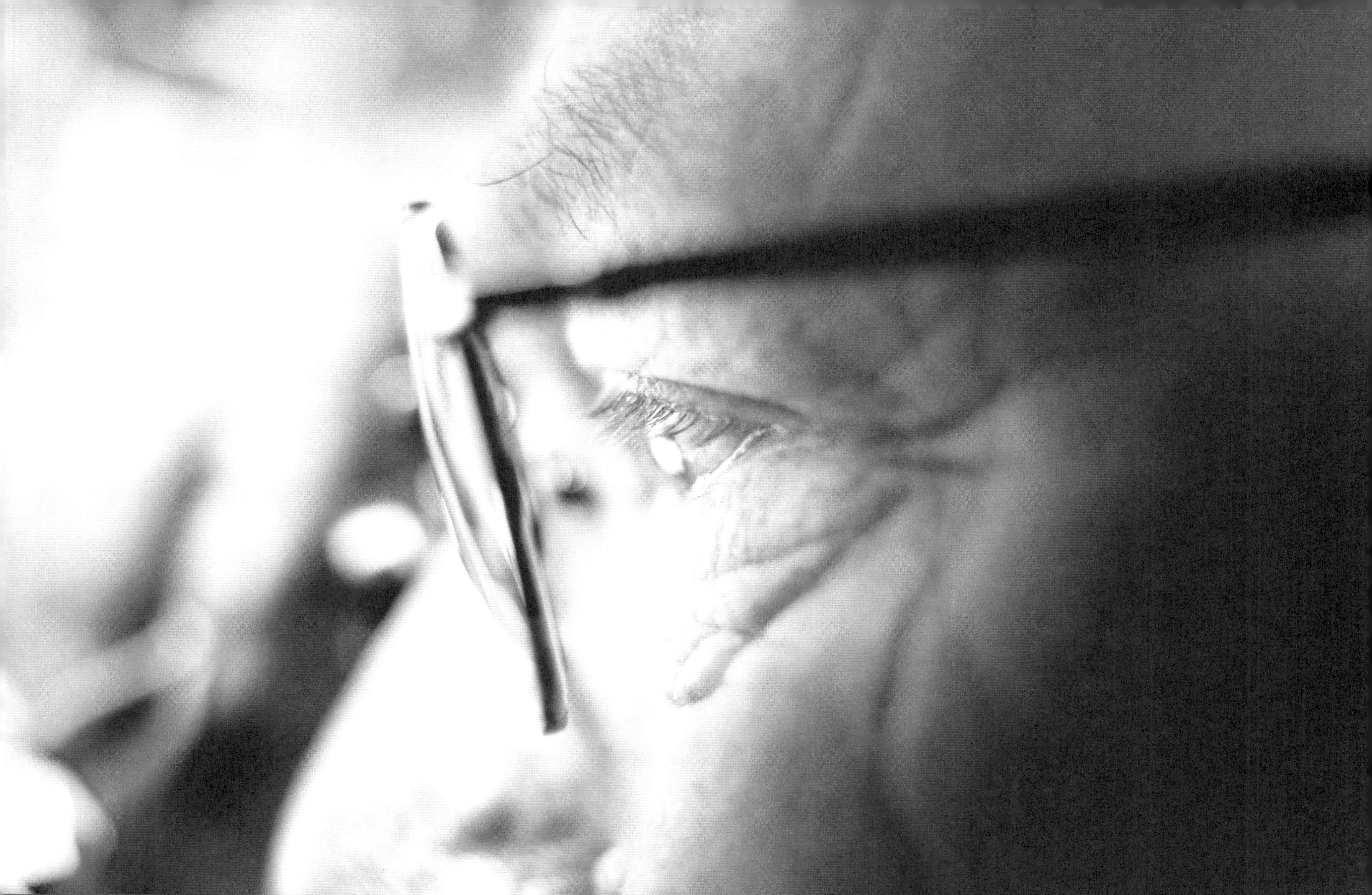

This “difference” is the mild-mannered, soft-spoken watchmaker’s way of dealing with his profession, consciously choosing to go down his own path, quietly and substantially. He is not so interested in seeing his name in lights, but rather in solving puzzles, his own or someone else’s.

Perhaps he gets this attitude from an earlier stage in his career, when he owned a repair shop in Zurich and often took on watches that his colleagues had simply refused to even attempt, simultaneously uttering the words, "That's no longer possible to fix." Gerber always found that to be a lazy way out, seeming to suggest to customers that watchmakers can no longer do what they have always been able to do. Gerber himself is the living confirmation that the modern era can indeed achieve what previous generations could–and more.

The year Gerber joined the A.H.C.I. he was asked to come to the following Basel Fair to exhibit in 1989, but he felt he needed a new project to bring along. So what did he do with only a couple of months to go before the fair opened? He created the world's smallest wooden clock–the movement of which was a mere 2.2 centimeters high–because he felt that a metal clock would take too long to construct. "I didn't set out to make the world's smallest clock–I just didn't feel like making a big one since I was making dollhouse clocks at the time and I simply didn't have the time to make one out of metal," he explains as his motivation for this feat.

P.+R. Gerber

Similarly, Gerber took on the Piguet/Muller/Gerber watch, generally known as Caliber 92, without stopping to think about the tasks set out for him by its then-owner, such as the addition of a flying tourbillon with diamond endstone, a split-second flyback chronograph (keeping many of the components visible, thus without a module), and power reserve indications on the twin spring barrels, along with turning the minute display into a jump minute counter. He just wanted to give it a go. Gerber relates that he often did this in his repair business, taking on difficult jobs from auction houses involving complicated, vintage masterpieces just to see what he could come up with.
"I really never knew in advance if I could [add the complications to Caliber 92]; I had never seen such a watch. I was just interested in finding out," he sheepishly admits today.

"There is not one invention in that watch that is mine. However, because of the conditions the owner imposed, I was not able to classically arrange the components, putting them where they fit rather than where I'd like to put them. It is not without reason that I needed eleven years to complete my changes to this watch."

"I simply like to practice my craft. I have to have it in my hands and see it, then I can say, 'That shape appealed to me on paper, but not really now,'" says this master of the quick prototype. In the blink of an eye he has made a case form or a bridge shape and tested it out, prototyping again and again until he has what his mind's eye was seeing–generally the last four-letter solution to the crossword puzzle or an object beyond most watchmakers' wildest reckonings.

True to form, Gerber plays down one of his latest inventions, a new escapement model that uses expanding forces rather than the usual compressing type, thus not entirely different from the reliable Swiss lever escapement, but new enough to make a difference to Gerber, who–again–just wanted to see if it would work. "On paper, I can of course prove that the absence of compressing forces in my escapement makes the component more precise," he says of his invention. "On paper, it naturally works better than the Swiss lever escapement, but I would never underscore that. I am just proud that my watches run just as well." It was neither precision nor oil that motivated Gerber, unlike many of his contemporaries: it was simply to see what he could do it. No more, no less.

Paul Gerber
CAL.33
Zürich

genau eine Linie

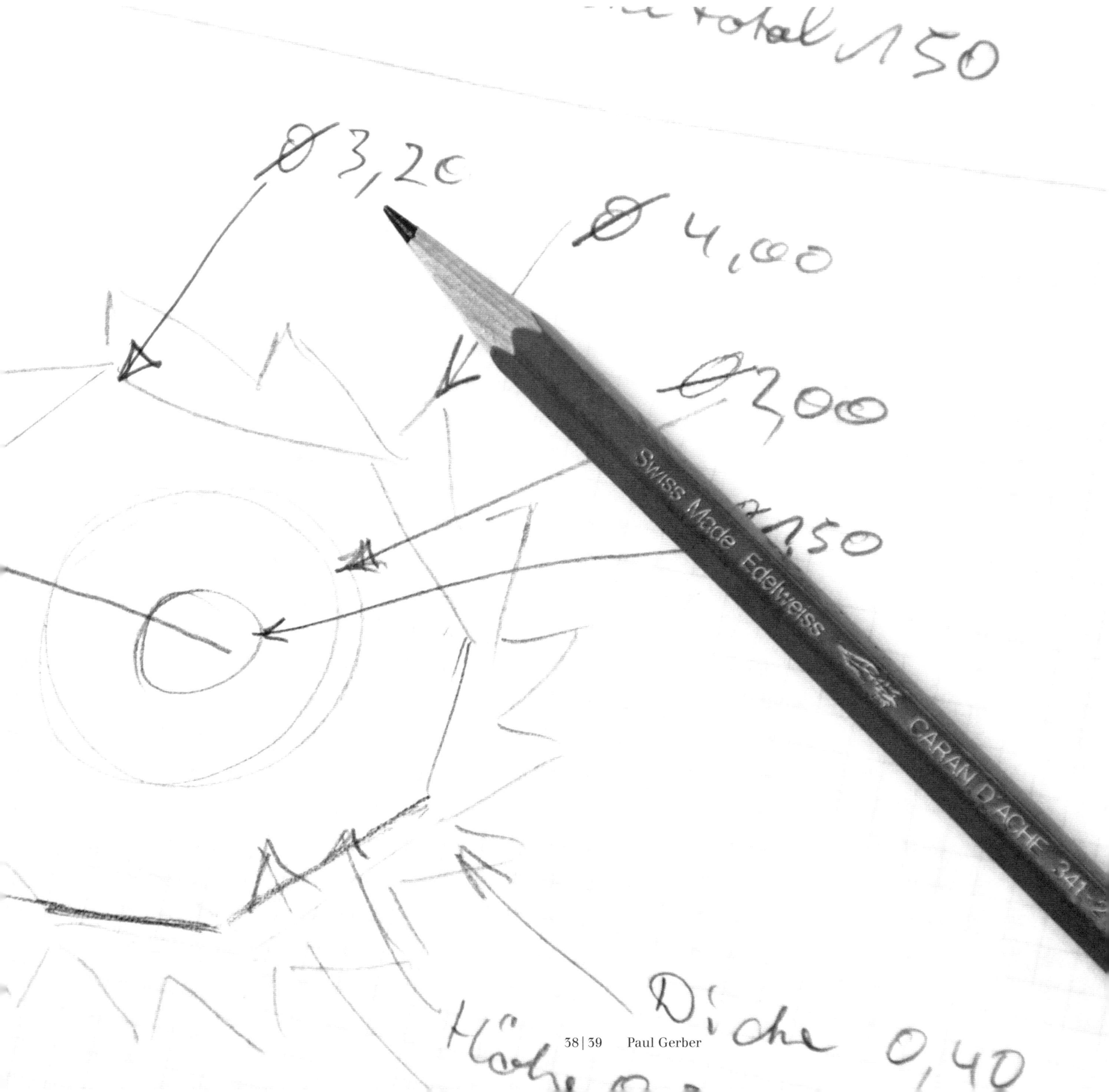
total 150
Ø 3,20
Ø 4,00
Ø 2,00
150
Swiss Made Edelweiss
CARAN D'ACHE
Dicke 0,40

"I try everything out; sometimes I just get intrigued as to whether I can do it or not. Who else tries to do things that haven't yet existed?"

Dr. Ludwig Oechslin

A master watchmaker who also has doctorates and other degrees in philosophy, astronomy, physics, and history? This is an unlikely combination and one that is certainly unrivaled within the watch industry. Though the bulk of Dr. Ludwig Oechslin's work is of an academic nature, his name is known in horological circles for its association with Ulysse Nardin, a company for which he has conceived many of the most unique wristwatches manufactured since the beginning of the mechanical Renaissance. An academic in every sense, Oechslin is one of the most diverse yet mysterious personalities to have ever graced the watch industry.

The result of Oechslin's labors is often poetry in motion: it's not about how quickly the escapement beats, or how fast he can finish developing a concept, but rather whether the result makes sense and meets his exacting standards. A new concept can and has taken him twenty years to bring to fruition, though some are completed in just a matter of months. His goal in conceiving a watch is met when he has achieved an authentic simplicity in something that is seemingly quite complex.

Oechslin burst onto the high watchmaking scene in the mid-1980s along with Ulysse Nardin and the rest of the mechanical renaissance. Although Oechslin had already been studying the humanities since 1972–when he began earning degrees in archaeology, ancient history, Greek, Latin, art history, and philosophy at the University of Basel–he wanted to work with his hands, but in a cerebral way. It was never Oechslin's intent to find fame in the watch industry. It, however, found him in the form of Rolf Schnyder, who showed up at master watchmaker Jörg Spöring's workshop in Lucerne one day in 1983 looking for someone able to make an astrolabe for the wrist. Schnyder had just purchased Ulysse Nardin.

Spöring happened to have taken on an extraordinarily academic, older apprentice from 1976 to 1984. Oechslin recalls, "Spöring was probably the only watchmaker who would have taken someone like me: having attended university and then going on to an apprenticeship at 24 was completely unorthodox. It was probably the thought of getting down to the seriousness of life itself that led me there, [to] something tangible, since with my other studies I couldn't really earn much money."

Schnyder got his wrist astrolabe from Spöring's able apprentice even though Oechslin was not really sure anyone would ever buy such a thing. Nevertheless, this timepiece–and the ensuing two completing Ulysse Nardin's Trilogy of Time–proved beyond a shadow of a doubt that Oechslin could funnel several areas of expertise into one flow of knowledge: the theoretical side of academics, methods he learned in his studies of archaeology, the examination and restoration of vintage movements, and the study of old models of the cosmos. "The historical studies led to my naturally having more of an overview of possible solutions already achieved," Oechslin recalls. "That in itself gave me ideas and led to creative new approaches. Without this historical knowledge, neither the Astrolabium nor the Planetarium nor who knows what else would have been created. It could only go hand in hand."

He continued to develop new concepts for Ulysse Nardin for many years alongside his academic pursuits, which included postdoctoral lecturing, writing treatises and other publications, research and professorial positions, and working for museums as curators and restaurateurs–and, of course, earning his master watchmaker title. The astrolabe and its two astronomical companions put both himself and Ulysse Nardin on the watchmaking map in the modern era, but Oechslin's practical side also led him to develop a GMT that could be easily moved forward and backward using two pushers, a "simple" perpetual calendar that can be set in both directions, and, of course, the astounding Freak–a study in unconventionality and innovation with its karussel tourbillon functioning as a hand. "The Planetarium is somewhat more reserved than its Trilogy counterparts, but in my opinion the most elegant in its solution," Oechslin says, looking back on the timepieces that made his reputation in the watch industry. "It was really fun to get to that point."

These days, Oechslin's considerable attention is mainly focused on outlets that include the Musée International d'Horlogerie in La Chaux-de-Fonds, where he is curator and director, and individual projects such as the continuing simplification of a perpetual calendar suited to a more moderately priced watch, some of whose profits flow into the museum. But of course, he won't stop there: he continues to develop various ideas, most of which are based on the ideals of simplicity and traditional watchmaking, and some of which continue to benefit Ulysse Nardin, which he still advises in a technical capacity. "I think it's fun until I have learned something from it, and then I hand it off to others," he quips.

It comes as no surprise to find out that Oechslin has dedicated a great deal of time and effort to researching the life and works of Jost Bürgi, a thinker born in 1552, apparently a renaissance man like himself, and a horologist about whom there is little information. Much of what is known about Bürgi today comes from Oechslin's study of the works the master left behind. "I think that Jost Bürgi was a central authoritative model, a figure from which one can start out again and again. When you look at him in total, he was probably one of the most innovative watchmakers of all time."

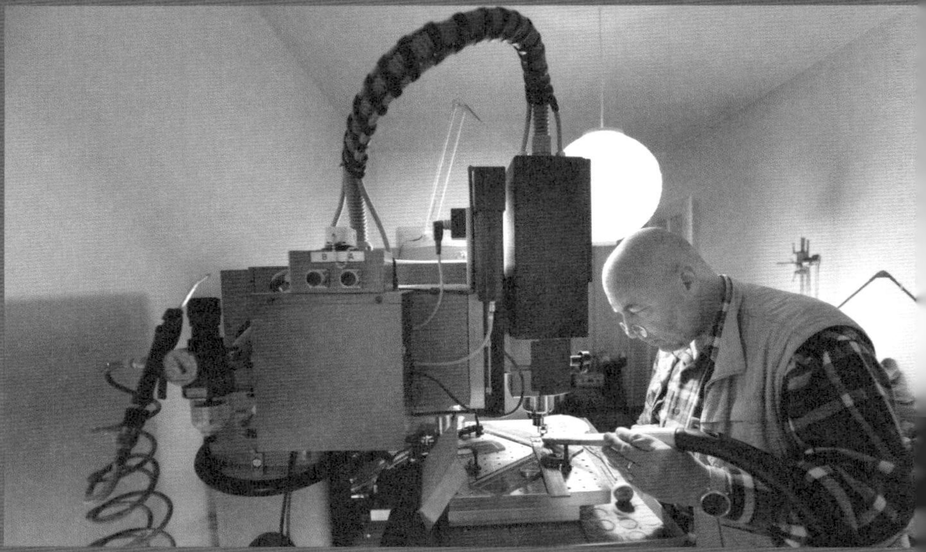

“I am uninterruptedly making things, working with my hands …that goes with the territory. I have to prove to myself that an idea can function. I am far too impatient; I want to see right away, so I do it myself.”

Though Oechslin obviously makes and enjoys timepieces that celebrate a relationship with humankind through culture, art and history, he doesn't hesitate to explain that his uppermost priority at this point in his busy life is family. "Naturally, sometimes [the family] does come in conflict with other things that need to be done. And it is work to learn how to deal with that and continue to do my things well, as well as possible, making sure the family doesn't come up too short."

François-Paul Journe

François-Paul Journe's workshop looks out over the busy streets of downtown Geneva, a sector of the city located in the heart of *haute horlogerie* that pulsates to the beat of its own rhythm. Those with watchmaking in their blood are attracted to its vitality like flies to sugar.

And contrary to his sometimes gruff exterior, Journe's head and heart are definitively channeled to his passion for watchmaking. Journe is not just a watchmaker, he is perhaps *the* watchmaker who has fueled collectors' thirst for products made by independent artisans and put this subgenre of modern watchmaking on the map. But at odds with his creative colleagues, Journe's interests lie not only in how to get his ideas to work, but in how to industrialize them so that his company is able to make more than just a few pieces of each; one of Journe's greatest goals has been to perfect his manufacturing processes–and to achieve this, he has needed to invest heavily in his workshop, machinery, and staff of skilled watchmakers.

Despite the appearances of his state-of-the-art workshop, his is an understanding of *haute horlogerie* that is rooted in the past. Respect for the greats of days gone by as well as both a determined belief in himself and an astonishing impatience have propelled Journe to the status of more than a mere watchmaker. These determining factors and his clever business practices have aided in driving him to a stature within the modern wristwatch industry that approaches the mythic.

F.P.JOURNE
Invenit et Fecit

Uncompromising in his search for the best possible quality in all respects, he is unbending when it comes to staying true to both himself and the finicky–sometimes unwritten–laws of high watchmaking. And for this reason independence at all costs is a must: he continuously reinvests profits into the verticalization of his timepieces, which simultaneously enables him to call projects to life that do not have to do directly with the propagation of his brand of wristwatches, but rather that of the culture of high watchmaking in a more holistic sense–including clocks, certainly less marketable today than a *montre bracelet*, and the organization of exhibitions in his attractive workshop foyer focusing on historical pieces. “I think that there is horology and then there are watches, clocks, and so on; horology isn’t necessarily ‘high’ or ‘low’...you have to start by asking, where is the horology at all?”

Undoubtedly, it has been the sheer complexity of Journe's timepieces that has catapulted him to the level he now graces. Upon creating his own Geneva-based brand after leaving the ranks of relatively unknown behind-the-scenes complication makers and restorers, the first "serial" watch he introduced was the Tourbillon à Remontoir Egalité, a constant force tourbillon that had not yet been attempted in wristwatch size to that point. Journe admits that he might have preferred to debut his brand with an elegant timepiece of historical inspiration à la 2005's Chronomètre Souverain. However, he did not have that luxury: he first needed to establish the collectibility of his brand and generate a word-of-mouth following. "This is logical," he explains, "because if you don't have the credibility, it is difficult to present a simple watch."

Journe has worked very hard to get where he is today. A rebellious type of kid, he was not one to conform to what others thought he should be in his early years in Marseille, where he was actually expelled from the watchmaking school he was attending as a last-ditch effort by his parents to get him educated. Taken on as an apprentice by an uncle in Paris specializing in restoration and with the same sort of unconventional attitude, Journe flourished at the most complicated types of projects he could find to occupy himself.

The characteristics he likely had as a boy probably continue to dominate his character in the present day: determination, perfectionism, confidence, and being ever-more demanding of his own abilities and of those around him. This emotional and physical investment has paid off for his little company in more ways than one: while creating a true investment platform for collectors of fine, unusual, and intelligent watchmaking, Journe has also created a platform for himself to achieve some of the most difficult and complicated horology available today–under constant evolution and evaluation, of course. When Journe decided that brass plates and bridges no longer sufficed for the level of watchmaking he wanted to create, for example, he simply began making them out of gold instead, a bold move in 2004 at a time when no other would have contemplated anything similar.

“I think, for one thing, that it is the poetical side that makes people love watchmaking and, for another, it is not something for the masses.”

F.P.JOURNE
Invenit et Fecit

Not every watchmaker has enough deference and belief in his or her own abilities to command the level of regard within an industry that Journe does. Journe respects his own point of view–a trait obvious in the way he has led his eponymous company since its founding in 1999–which more often than not represents the fundamental values of *haute horlogerie* as they have been practiced for centuries. And while Journe accepts that there is space for many in our modern world of high watchmaking, he does not necessarily accept that everything called *haute horlogerie* is exactly that.

Smart business practices that include fascinating timepieces and a personal, hands-on concept make Journe SA very unlike other similarly sized companies. At the same time, a certain audacity maintains a chain of surprising complications: for example, housing one of the world's most complicated timepieces in steel, an element that derives from a technical background rather than one professing perceptions of luxury. Journe feels no need to explain himself to anyone; it is his opinion–backed up by research and history–that a Sonnerie simply carries most beautifully through steel.

"Right now, it is more its degree of artistry that ensures that horology will continue to exist, because from a scientific point of view it is an industry that is completely inconsequential," Journe explains in his lilting French. "Horology no longer has a reason for existing today. It isn't quite normal that people love mechanical watches in the twenty-first century, which is an era of information."

Kenji Shiohara

In the heart of the most technologically advanced landscape on the earth, that of Japan, sits a wellspring of artistic horological creation.

Surprisingly, in the midst of all that forward momentum there is one who would take a step back in time to learn from the masters, integrating what they knew into the innovations emanating from one of the most advanced timepiece manufacturers in the world. Kenji Shiohara's destiny thus entwines with the culture of Seiko, the oldest watch company in Japan and one of the largest in the world.

Since the early 1970s, when Seiko proudly struck out into the quartz era it had worked so hard to create, efficiency and automated production have been the concern's declared goals, and craftsmen no longer needed to be skilled in traditional watchmaking.

Shiohara, one of the last employees at his factory to acquire mechanical watchmaking expertise, excelled, becoming a First Class Skilled Watch Artisan, a Japanese national qualification that corresponds to the established title of master watchmaker, and winning a gold medal in the now-defunct World Skill Competition. And yet, Shiohara always knew he could do more to express his creativity.

Then, as coincidence would have it, he got his chance.

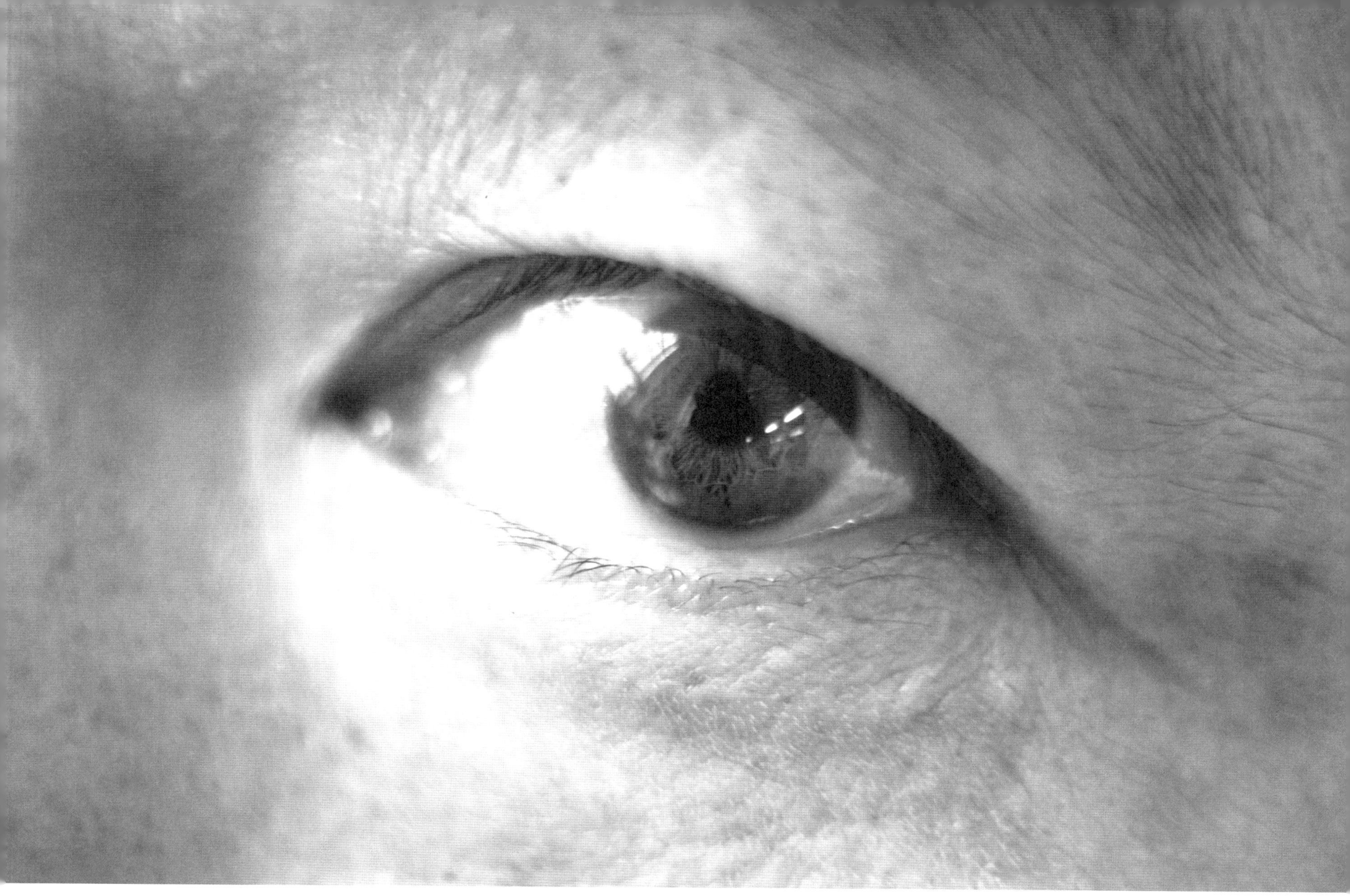

In 1999, Shiohara was asked by the then-president of Seiko to repair his personal watch, a mechanical timepiece a mere 1.2 mm high. This was a feat no one else at the concern would probably have been able to accomplish. Working with the filigreed hand-wound watch got Shiohara's creative juices flowing, and, in a very forward move considering the norms of Japanese culture, he voiced to his superior his personal interest in creating a sub-department for inventive watch-making within the great Shiojiri factory.

CREDOR
MARCHE
SONNERIE
SONNERIE
SILENCE
ORIGINAL

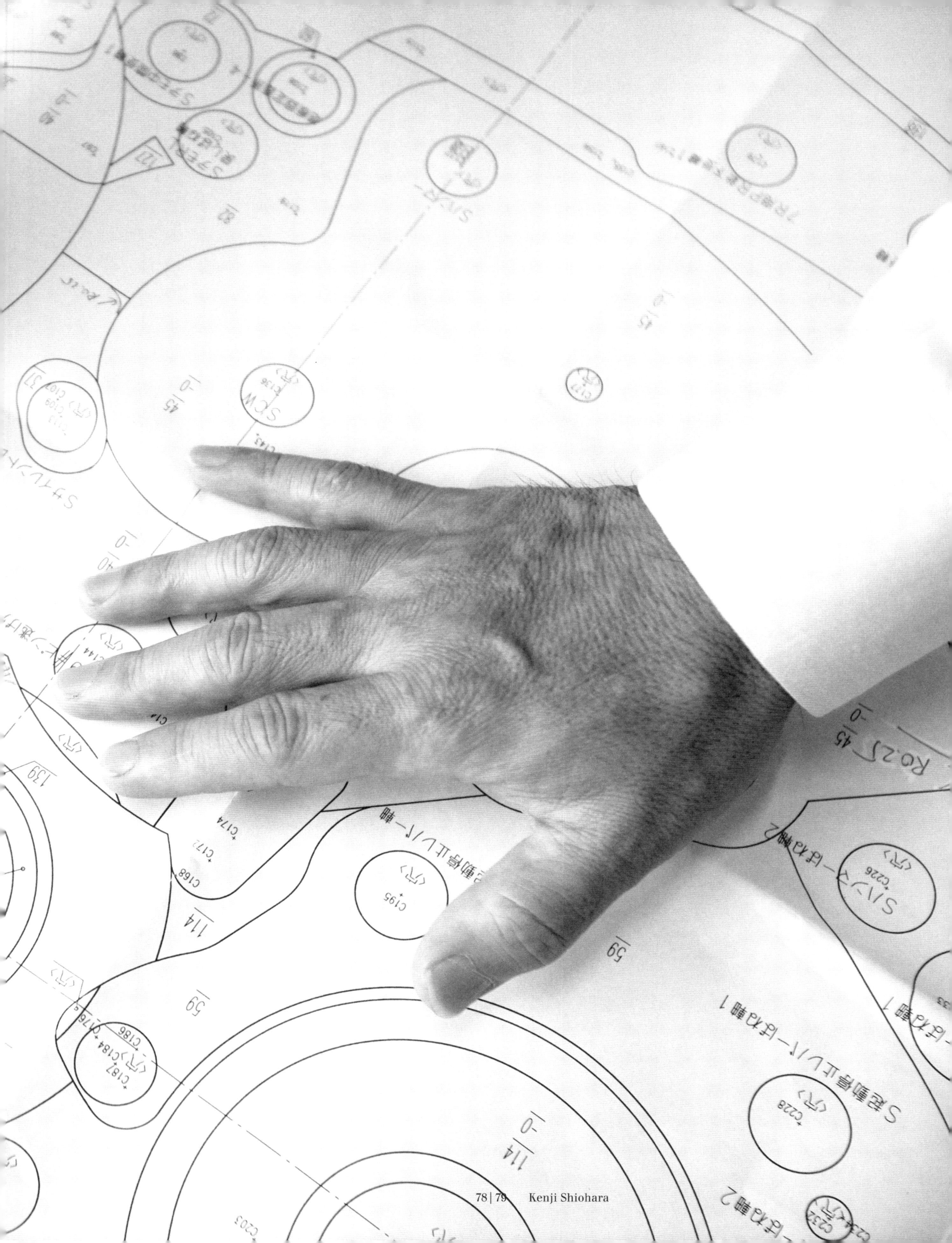

Not long after the establishment of that department, the Micro Artist Studio, in 2000, Shiohara saw an interview with Swiss master watchmaker Philippe Dufour on television. Shiohara was already fascinated by Swiss watchmaking, and this TV program delved into the small universe of independent watchmakers so greatly respected in Japan.

It was during this interview that Shiohara learned that traditional Swiss watchmaking can extend the life span of a wristwatch. This and other things Dufour had to say made Shiohara feel that they were kindred spirits. Although separated by thousands of miles, cultures, and very different languages–barriers that might well have seemed insurmountable to anyone other than these dedicated proponents of high watchmaking–Shiohara and Dufour managed to arrange a chance to exchange thoughts; a meeting of minds that turned out to be a distinct enrichment to the world of watchmaking.

Shiohara learned of Dufour's attitude and way of teaching others watchmaking in a manner not limited to just one culture or mode of thinking, and his obvious pride in passing on the methods of watchmaking to following generations struck a similar chord in the Japanese master.

The independent, ten-man-strong department that Shiohara had called into being within the gargantuan Seiko concern continues to learn from all types of sources, moving beyond imitation to synthesize the myriad inputs into unique Japanese cultural goods. While making these original timepieces, the Micro Artists also embrace the goal of handing the skills they have acquired down to the next generation.

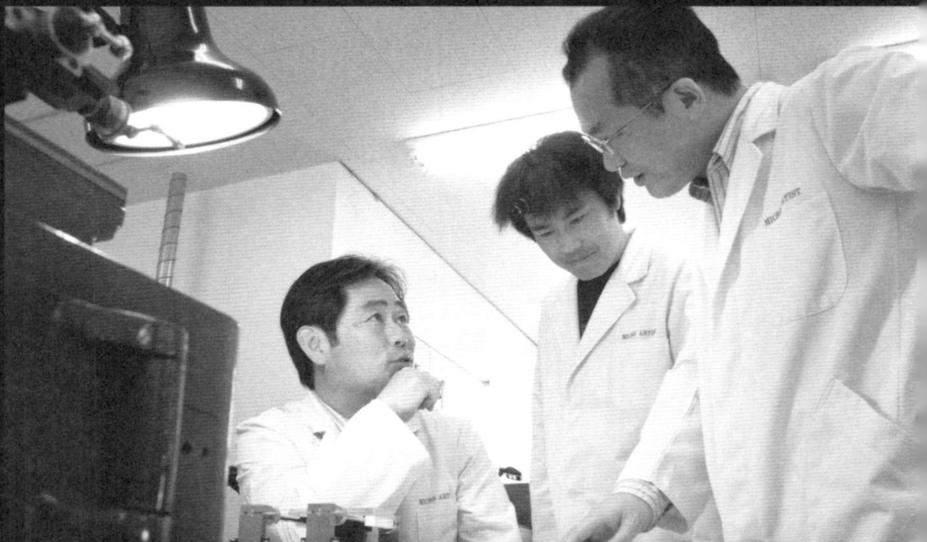

The masterful timepieces that have emerged combine *haute horlogerie* production and finishing techniques with the high-tech reliability of the Seiko Spring Drive and are characterized by Shiohara's basic philosophy: to create a good watch that will last for several lifetimes. "Artistic finish is one way to make a quality watch. Handing down skills includes handing down decorative techniques like engraving and lacquering," Shiohara goes on to say of his department's philosophy. The Micro Artists strive to achieve high-end watches that incorporate Japanese design and cultural components with precision horological techniques that have been developed for centuries.

The Credor Sonnerie, in particular, enraptures the imagination as a visible product of Japanese sensibilities and the Micro Artists' finely tuned goals. "Before clocks and watches, in Japan it was the temple bells that provided a reckoning of time. The sound of a gong has a special resonance for a Japanese person; the temple gong provides a lingering tone that can cross a great distance. A church bell is noisy to us; we feel more comfortable hearing the shorter striking of the temple gong," Shiohara explains.

He grew up in the area surrounding the Shiojiri factory; his family had no watches, simply marking time by the tones from the nearby temple. Thus, Shiohara never felt pressured by time, and it is this feeling that he wanted to reproduce. Shiohara and his team developed a component for the Sonnerie that not only produces a similar sound but also bears the shape of a Japanese orin bell.

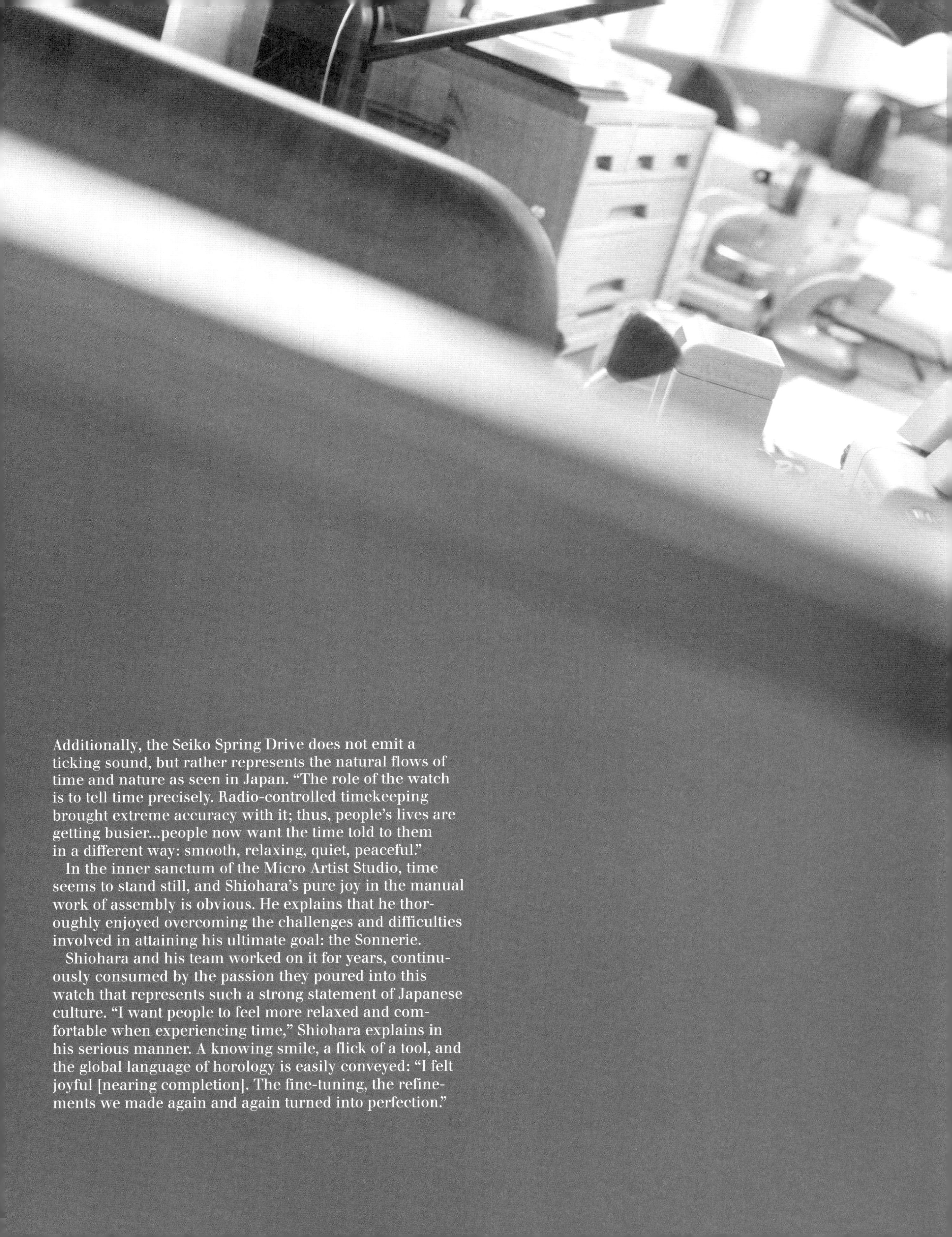

Additionally, the Seiko Spring Drive does not emit a ticking sound, but rather represents the natural flows of time and nature as seen in Japan. "The role of the watch is to tell time precisely. Radio-controlled timekeeping brought extreme accuracy with it; thus, people's lives are getting busier...people now want the time told to them in a different way: smooth, relaxing, quiet, peaceful."

In the inner sanctum of the Micro Artist Studio, time seems to stand still, and Shiohara's pure joy in the manual work of assembly is obvious. He explains that he thoroughly enjoyed overcoming the challenges and difficulties involved in attaining his ultimate goal: the Sonnerie.

Shiohara and his team worked on it for years, continuously consumed by the passion they poured into this watch that represents such a strong statement of Japanese culture. "I want people to feel more relaxed and comfortable when experiencing time," Shiohara explains in his serious manner. A knowing smile, a flick of a tool, and the global language of horology is easily conveyed: "I felt joyful [nearing completion]. The fine-tuning, the refinements we made again and again turned into perfection."

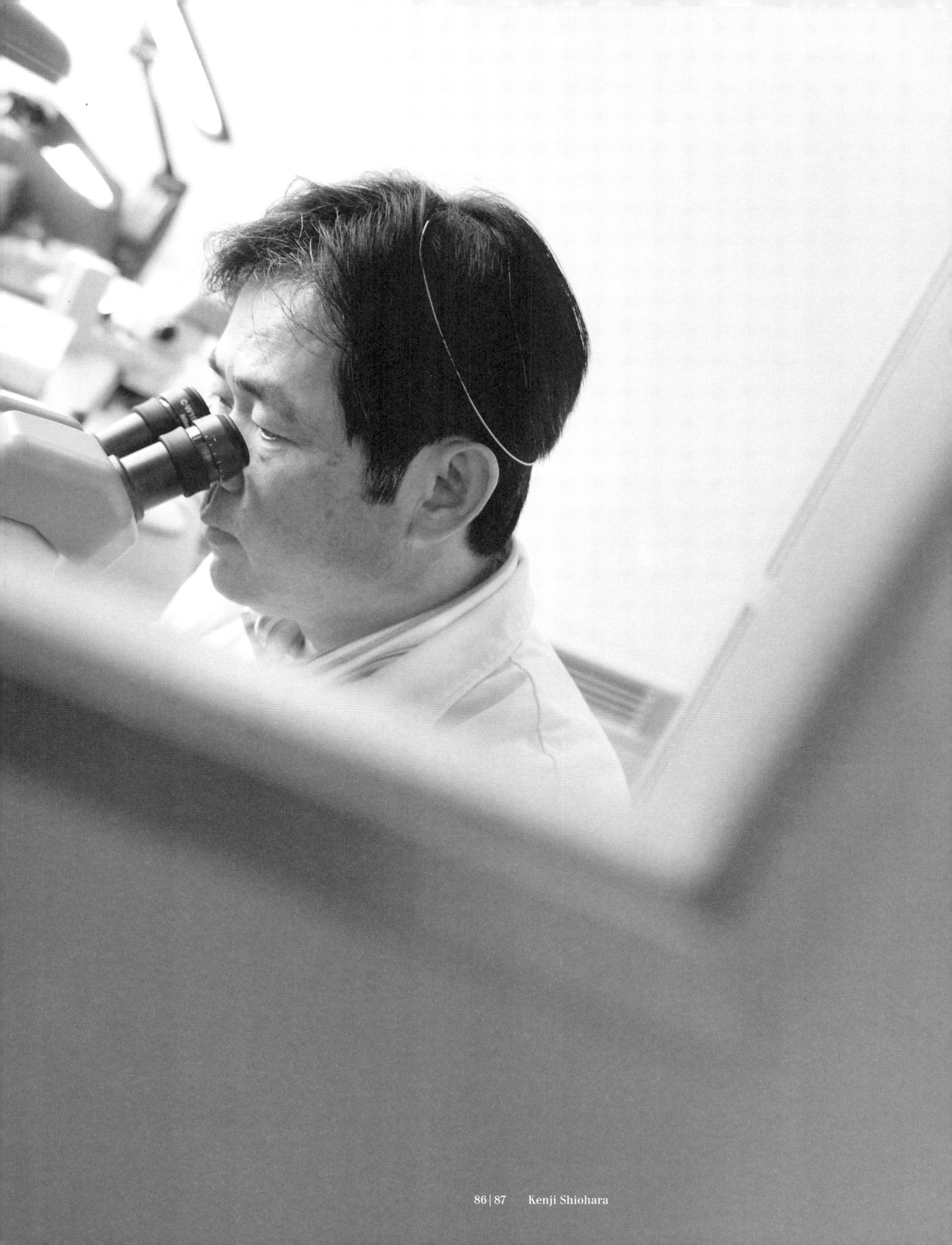

押
PUSH
A1-5

“I appreciate my predecessors’ wisdom and love to improve on that knowledge. I also have the responsibility to hand down that knowledge to the coming generations.”

Kari Voutilainen

“This is my universe; my whole life turns around watches,” declares Finnish master watchmaker Kari Voutilainen, at his home in Switzerland’s Val-de-Travers, a region blessed with terrestrial magnificence.

Voutilainen, always enamored of his own homeland's natural beauty, took quickly to the solitude and alpine splendor of Môtiers, where he has now lived for almost two decades. Passionate about the craft he calls both his hobby and his vocation, he traveled to Switzerland to complete his education and was quickly hired by Michel Parmigiani, with whom he successfully executed complicated restorations of vintage timepieces.

Voutilainen has spent his career working on horological œuvres at home as well as "at work." By 1994 he had spent 2,000 hours creating a pocket watch tourbillon that included a handmade movement, case and dial, all done in the evenings and on weekends. The next project to occupy his leisure time, between 1995 and 2000, was a classically styled marine chronometer.

Industrious by nature, the modest watchmaker nevertheless discovered that this way of developing ideas was akin to holding two full-time jobs, so he decided to allow his passion to determine his fate by founding a workshop of his own in 2002. "It was easy to strike out on my own," he smiles gently now at the thought, though the initial stages could not have been quite as effortless as he makes it sound: to get the financial ball rolling he started out by prototyping and designing movements for others and even translated technical manuals using his outstanding language abilities.

His brand of watchmaking is a distinct continuation of traditional style. "I'm not a designer, I'm a watchmaker," he states, explaining that he begins a timepiece with the movement rather than the watch's appearance, although "the result will always have my gold thread running through it–in the end it always looks the way I want it to." And his way is elegant, never overbearing, very detailed–faultlessly mirroring the Finn's own personality.

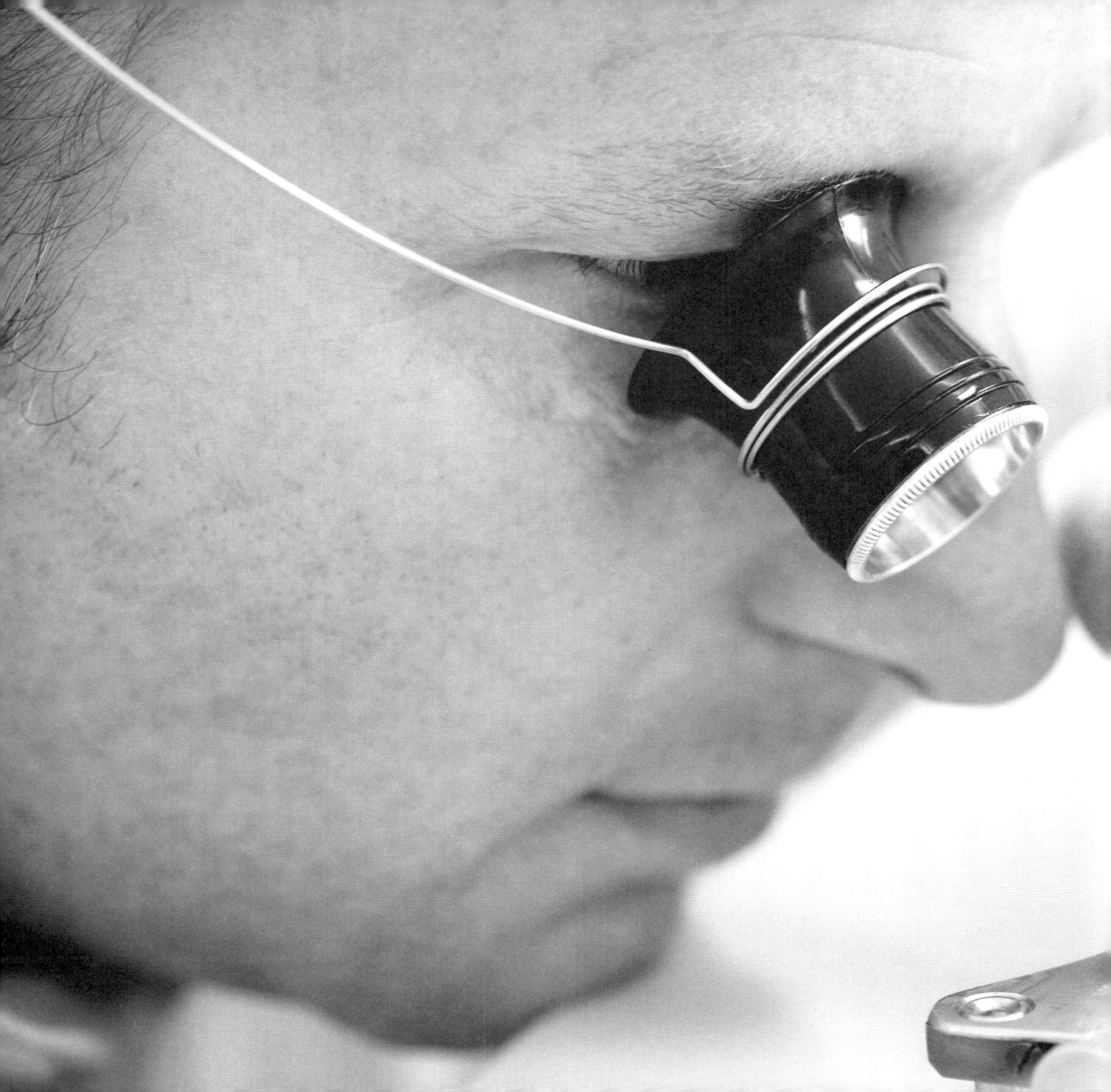

Voutilainen's basic philosophy comprises respect for the work and the components–and perfection in completing both. Nowhere is this as obvious as it is in the amount of time and effort he and his small team spend on the movements. Voutilainen explains that component manufacture is done rather quickly with the assembly being only about 5 percent of the total time dedicated to a movement. Finishing, decoration, and post-production processes account for the lion's share of the duration. "Finishing just takes time," he explains in his patient manner. "All the other work we do on our watches is much more important than the actual assembly."

Most of Voutilainen's masterpieces are frosted, a highly difficult finishing method known in French as *grainage* for flat parts to make them harder and more resistant to corrosion. There are very few manufacturers currently utilizing the technique. "You need experience to do this well. You cannot make a mistake," Voutilainen explains as he illustrates the process completed with a bent steel brush. "Aesthetics are just as important as the technology," he philosophizes, "and it is tradition that dictates what is considered high-quality decoration."

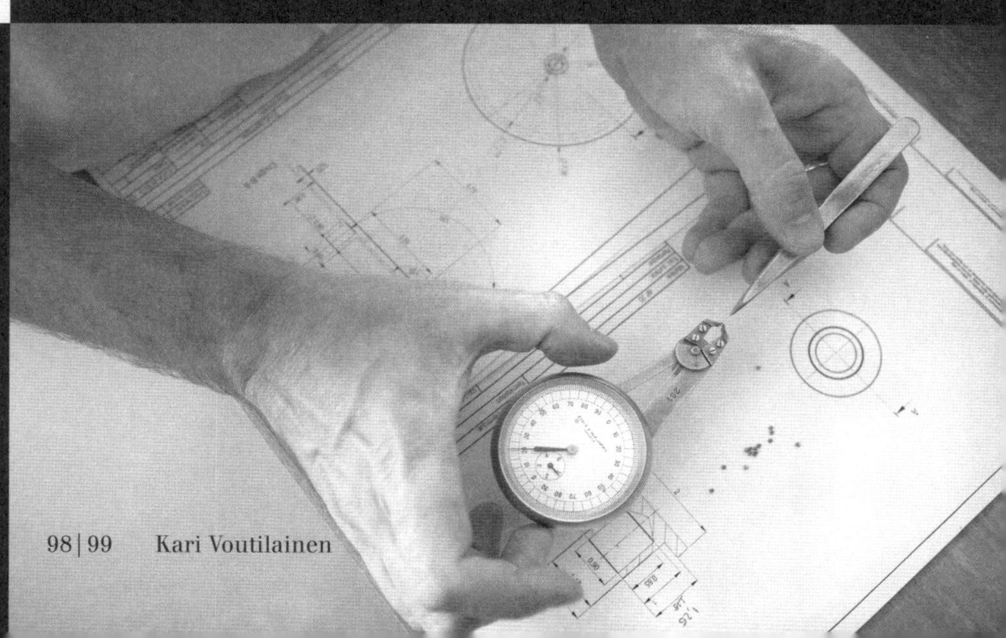

"A watch should be worn.
People should enjoy them,
that's what they're made for."

While clearly influenced by his extensive background in restoration, Voutilainen above all loves clean, clear shapes, some of which are inspired by industrial design of the 1950s and '60s; his look remains elegant, never approaching anything remotely trendy. Voutilainen loves working with applied elements on his dials, which "give both a three-dimensional effect and strength." He never uses precious metal plating, even here: what looks like gold is always gold, both complementing and embodying the substantiality of his achievements.

"Watchmaking is so personal, that's why I sign the watches myself, like a painter who signs his painting," which explains why he loves working with the collectors who sometimes travel great distances to purchase the timepieces he makes. He greatly appreciates the direct communication between client and watchmaker. What you see is what you get: "I'm making my own history. My clients meet and deal with the living person," 2007's winner of the Grand Prix d'Horlogerie for the best men's watch explains.

Another element also makes his watchmaking art personal: his case backs are, and always will be, transparent regardless of the mood of the times. “Most of my clients want to see the movement,” he states. Both parties enjoy the visibility of the perfectly finished mechanics at work—which is why he will never close off his mechanical beauties inside an iron core to make them even more antimagnetic than he already takes pains to do.

The only fly in the ointment for Voutilainen is that time is always so short. The core element his livelihood, and indeed his life, revolves around is his only fetter. “I have so many ideas…it is a problem of time. I love creating. Realizing the evolution is the best, getting it to work. It’s really rewarding.”

Roue
De
Arbre de barillet
Tambour de Barillet
Couvercle de Barillet
Bascule de correcteur date
Roue unités date
Roue de remontage 2
Roue de remontage 1

Vianney Halter

To look upon his horological œuvres is to hear a bird sing: the observer without a doubt understands all at once that he knows how to take all the notes available to a musician and put them in new order–making a song that has never been sung before.

As a bird sings what he feels, Vianney Halter creates what he dreams. This is far more than mechanics, or the understanding of these mechanics; his watches are literally the tangible reality of what is inside him.

When Halter created the Antiqua model in 1998, he embodied his own vision of perpetual time, giving his perception of this immortal element a shape and a home. Several years later, after Halter had founded his own *manufacture*–named in honor of his favorite watchmaker, Antide Janvier–he hired a movement designer to realize his designs tangibly, and this new computer specialist discovered that the Antiqua should by rights not work. Halter's innate understanding of watchmaking was the overriding factor in the success of the Antiqua's caliber, allowing it to perform the task its master had set out for it. Logic, however, appears to disagree. Halter chalks up this seeming paradox to experience at the bench, something a computer does not have.

In the same vein, Halter compares the apples and oranges of mass-produced watches to handmade timepieces: "Even though we are making the same sort of end product, we have different ways of achieving it. These two ways of manufacture are complicated forms without any real connection between them. Some elements and aspects of these products are similar, but if you go deeper into the story of each, you see they're completely different. It's impossible that these can be called the same machine."

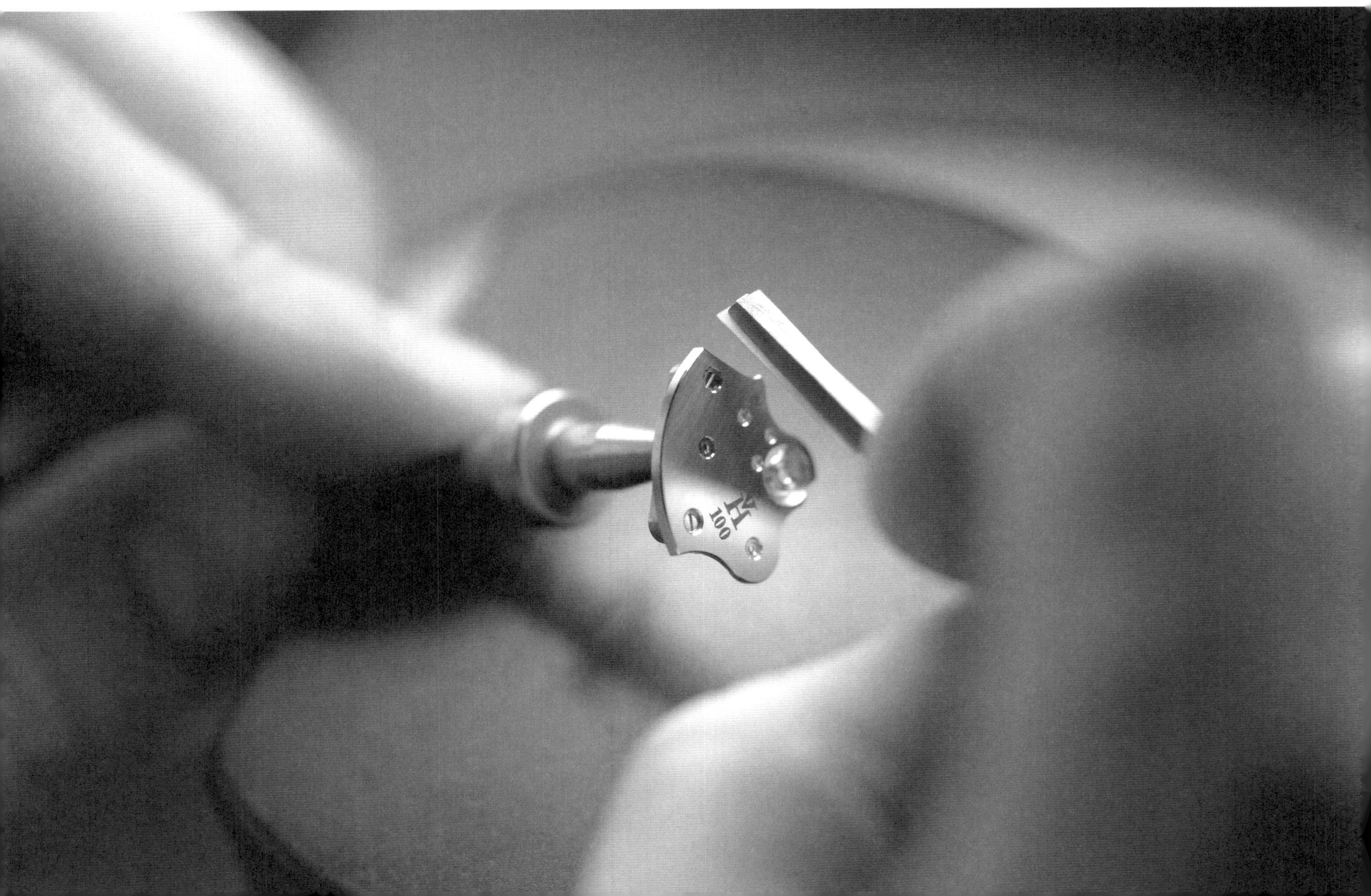

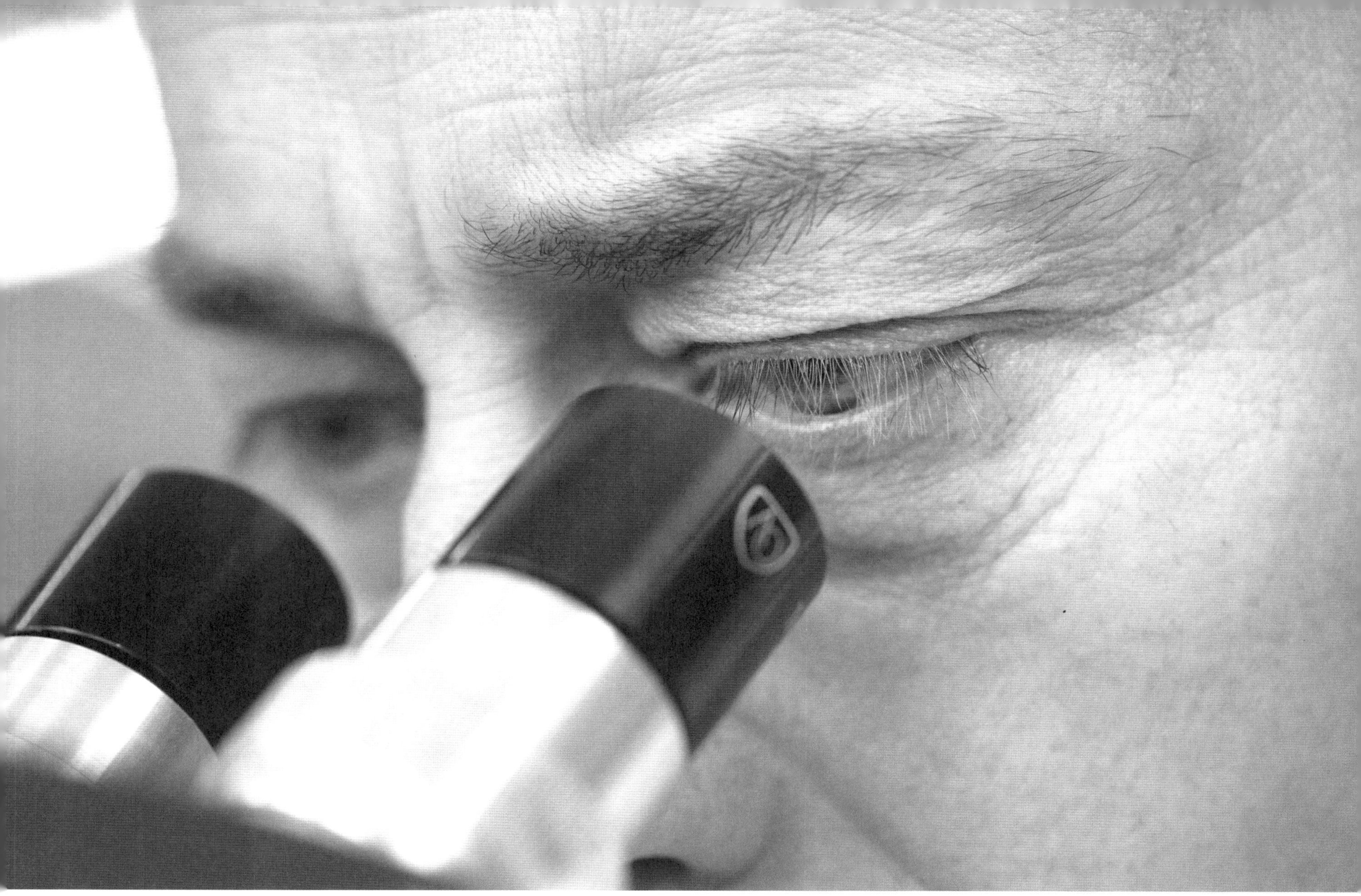

The pervicacious watchmaker enjoys his position as an independent maker in the watch industry, relishing a new style of watchmaking that he helped set into motion. "The real watchmaker now works more and more alone and less and less for big factories; the real watchmaker follows a way that's different from this traditional production method," he explains in his excitable manner. Halter considers his experience one of his biggest assets and what allows him to realize what he sees in his mind in the physical realm. His firsthand knowledge comes from restoration and the practical work he quasi-legendarily completed behind the scenes of the Swiss watchmaking industry.

"We don't live like the people you find in American movies, something like *The Matrix*, where you can immediately dump the knowledge you need to pilot an aircraft right into the brain. We work with it [to gain the experience]," he explains.

For Halter the watch is emotion. Plain and simple, there are no other whys or wherefores. Halter does what he does because he loves it; he is not about branding or getting rich, he simply wants to make watches. “I follow this work because it’s a language for me,” the charming Frenchman explains with wildly gesticulating hands and facial features to match the sparkle in his blue eyes.

He goes on to explain that although he has a tough time conversing in English, he and Roger Smith, for example, understand each other better than most others, speaking the horological language that they have both spent so much of their lives perfecting. It is a language without words, a language of shapes and textures, a language with its own non-computer-generated syntax.

"And maybe, just maybe, other people, not [the generation] after me but later, *plus tard*, will come across this base and understand what I am explaining here, and, maybe, they will also talk through their work, using the same language, and continue to talk with different people around [them]."

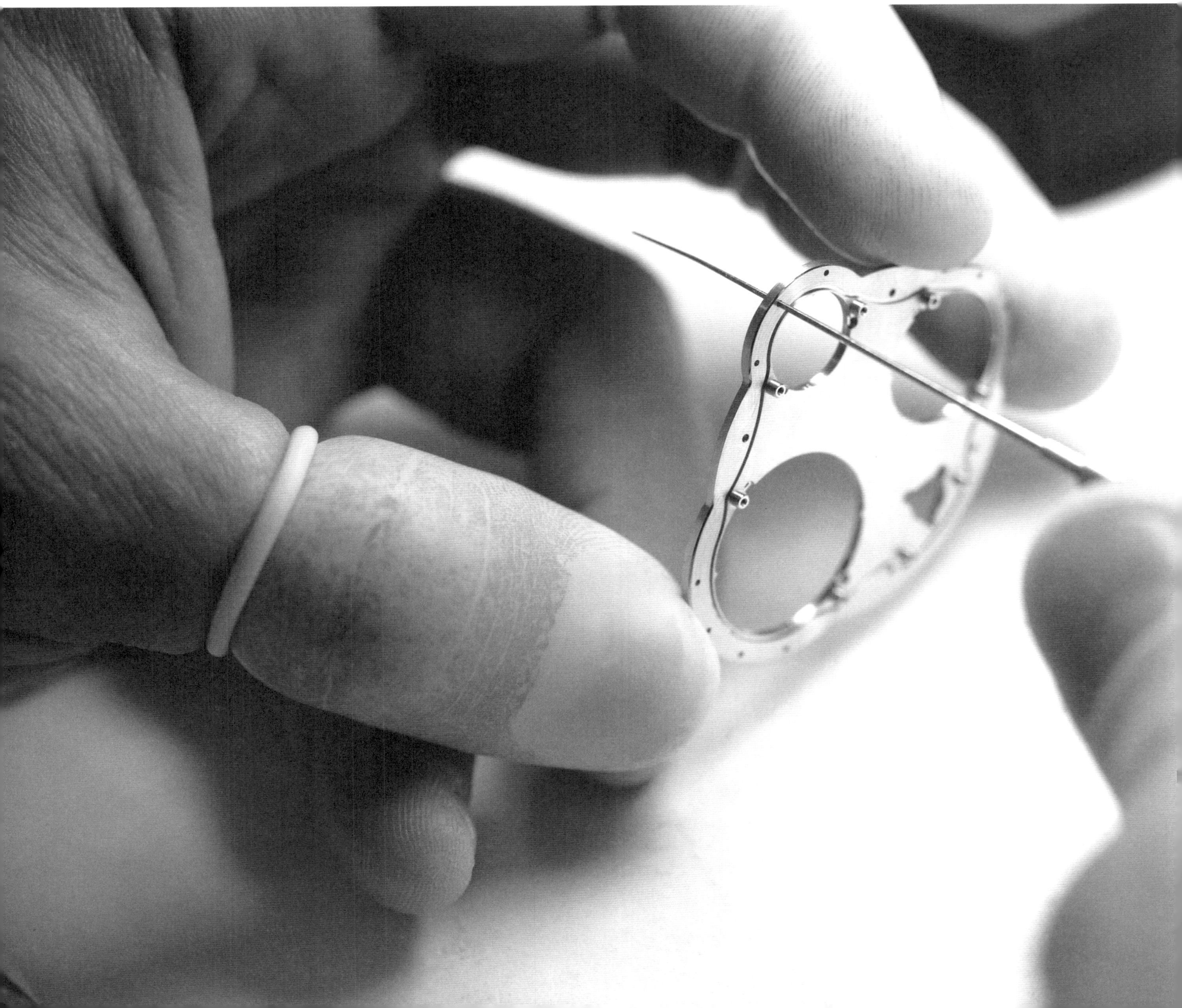

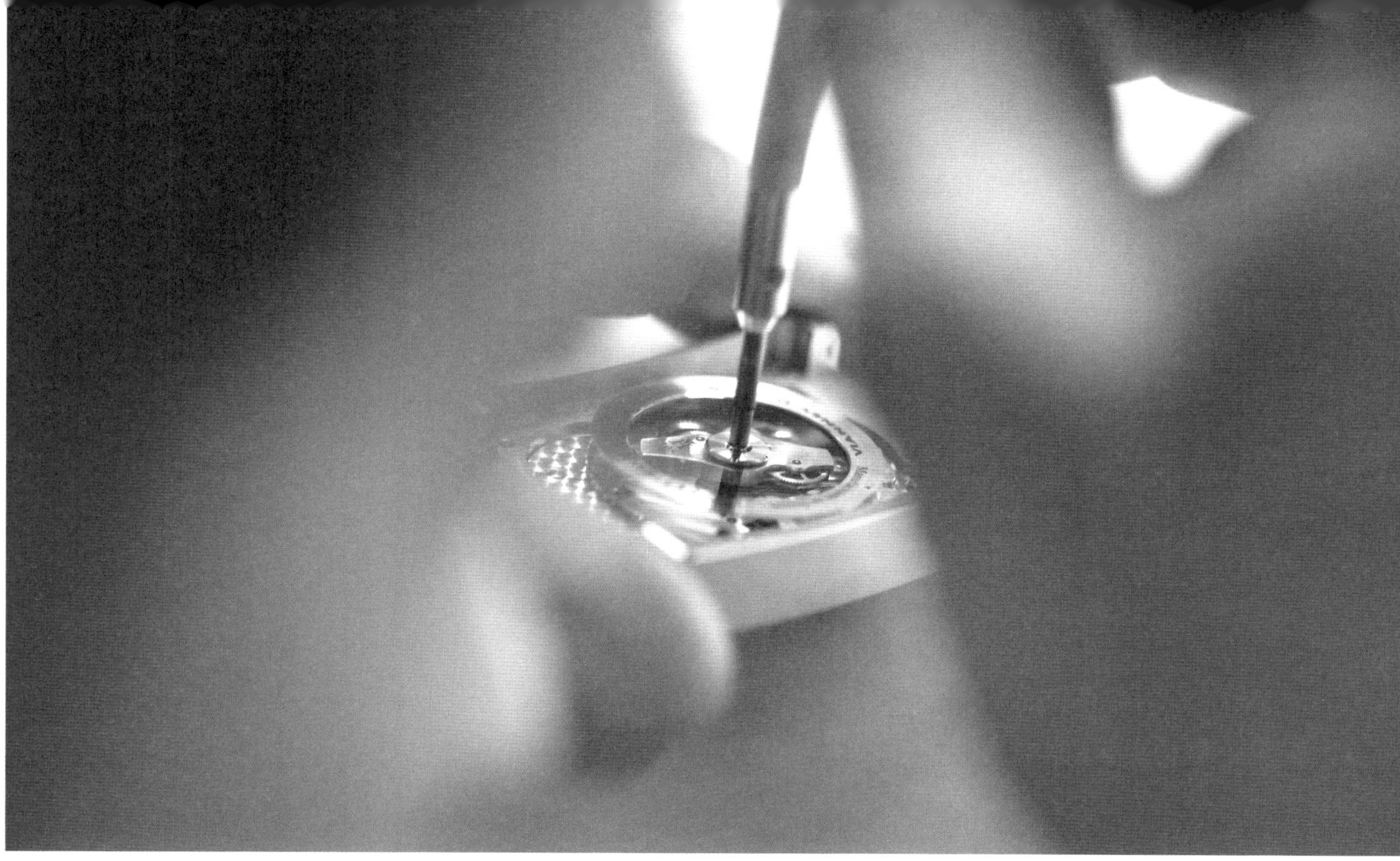

Halter decrees that his watches must be meaningful, telling a story of their own. Though it may not seem like it, Halter's work is based in heritage and tradition, and his timepieces must exude a certain equilibrium.

"I feel that a watch is balanced when, regarding it, you cannot add something and you cannot remove something without changing the balance. For example, my Classic Janvier includes the equation of time and the moon phase because it was dedicated to Antide Janvier, an astronomical clockmaker. Of course, we could add a tourbillon. But what would be the meaning of this tourbillon?"

The most important thing for Halter, however, is to live each day to the fullest, making it interesting and strong. "I don't want to have a day when I feel it's not me."

Halter compares life to flying a plane, another of his passions. "Sometimes problems show up [in the cockpit], but the target is to decide to have control over the direction you choose and not let the machine choose the direction for you."

This sounds irresistibly like he is talking about his own Jules Verne style of watchmaking, though he quickly veers back on course. "I like piloting because I am the one who is in control, not the machine. Nobody's around to give you advice. If you take time for advice, you crash."

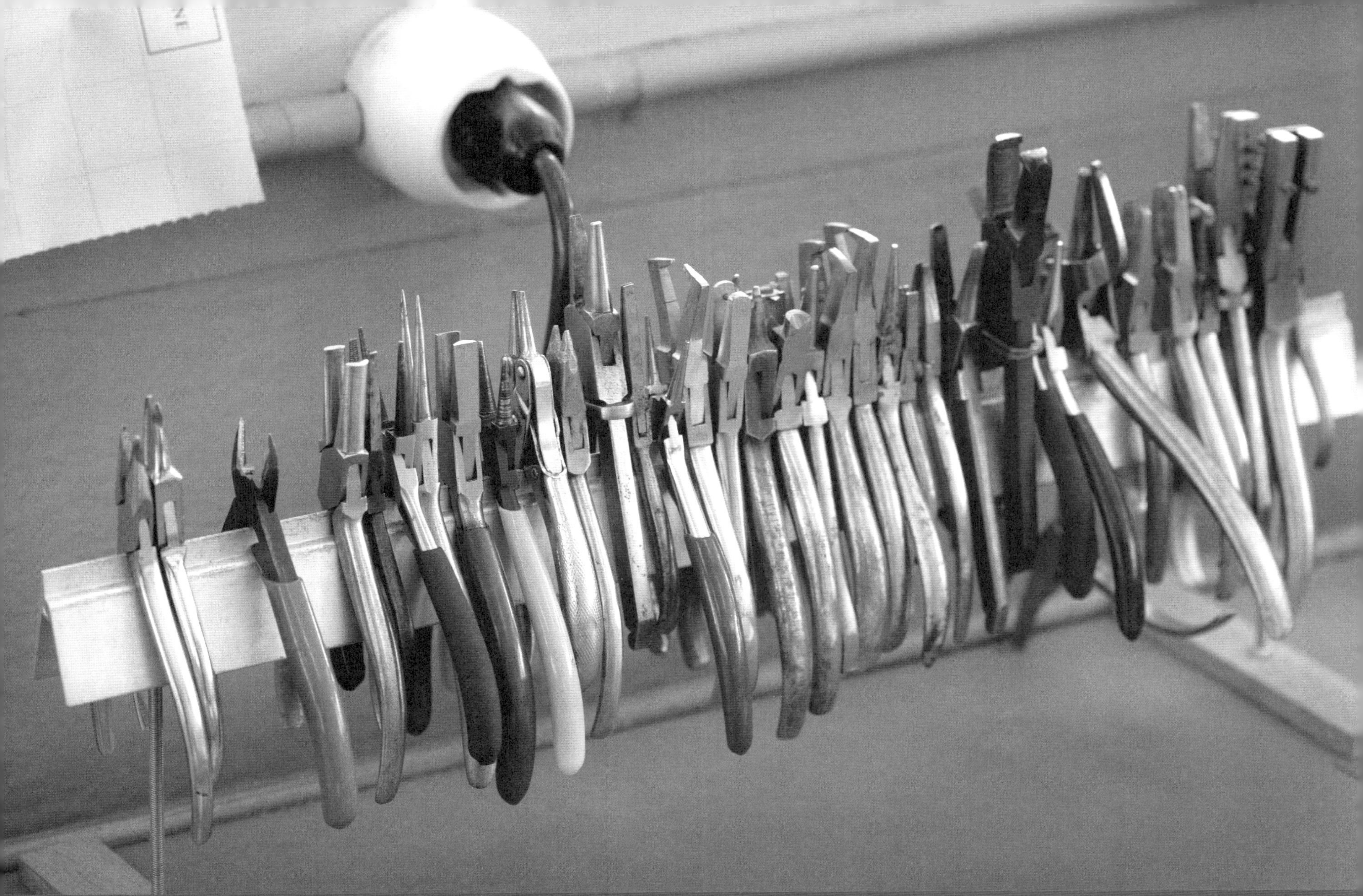

Often touted as a genius in the world of horological mechanics, spending an afternoon with Halter is like spending an afternoon with your alter ego–the one that wants to live out all its fantasies that aren't allowed. Here all is allowed. "Sometimes you don't know why, but a certain energy or instinctive feel for heritage pushes you in this way, and you arrive [where you wanted to go]."

It is easy to draw parallels between Halter's inspirations and products and artists working in other fields. "A painter has a vision, and color, and he pushes, and after he arrives ...for many years, people see this with admiration, but nobody understands it. 'Oh,' they say, 'it's beautiful, yes, yes, but too crazy for people to understand.'"

"It's very important the next day [that I am] exactly what I was yesterday, and what I want to be tomorrow, next week, next month, next year. I don't get too old too quickly, I think, because I remember my youth, my dreams, my impressions, my feelings, my sensations, and each new day I look to find them again."

When standing next to Beat Haldimann, the first thing any observant person should notice is that you can hear the watch he wears loud and clear–it's an early prototype of his H1 in a non-precious case. These kinds of sounds are very important to him: tick, tick, tick on all the walls; tick, tick, tick in every room of his house and workshop; tick, tick, tick on his wrist. There is peaceful, rhythmic ticking everywhere, at times combined with the melodic song emitted by his tourbillons.

Beat Haldimann

This perfectionist element extends throughout Haldimann's entire being. Audibly and optically, everything he undertakes is connected by a distinct golden thread. He lives and works in a mansion built in 1907 in the German Secession style. Consequently and consistently, Haldimann has filled the house with furniture, tools, and decorative objects originating in the same artistic period. And this is the same sort of process that permeates his horology as well. Everything has its place, nothing appears orphaned: all is as it should be, in harmony with the master watchmaker's grand plan, even if it takes years for him to get there.

Haldimann founded his company in 1991, but it was only eleven years later that he acquired the estate called Nussbühl, "room for room" as he describes it, renting workspace within the grand house as it became available, then finally purchasing the whole thing in 2002. He relates that it has taken him the past eighteen years to get his company to the point where he feels that it is now running efficiently and much as he had imagined it.

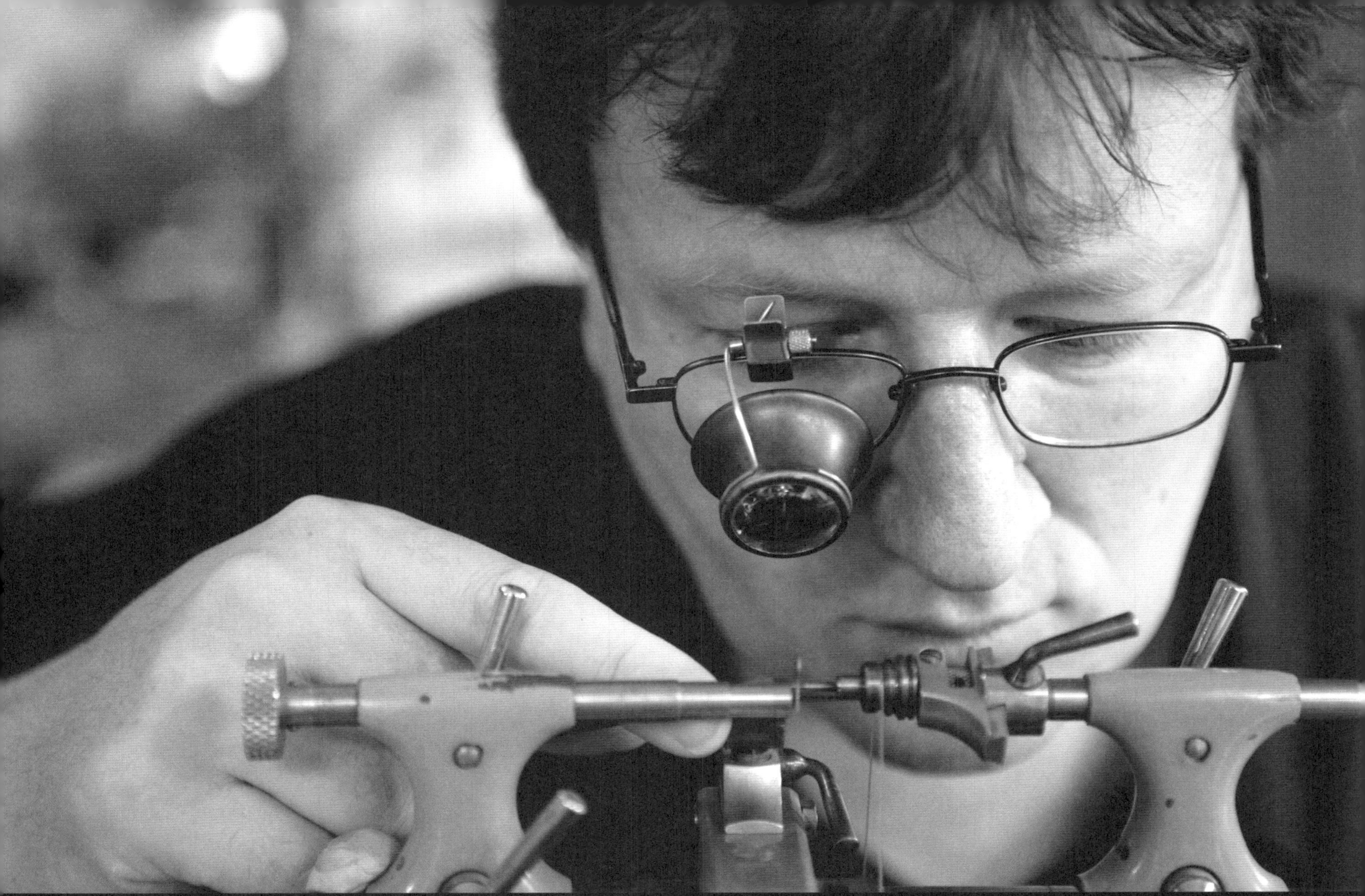

His watches are also now what he has long envisioned. "My philosophy is that of reduction. I think it is brilliant if you can't take anything else away from [a watch]–and not when you put more into it," Haldimann explains, before describing how people first gravitate to the central tourbillon on his timepieces before actually noting the time–the main function almost becoming an afterthought. This is because Haldimann consciously puts so much emphasis on the wearable, kinetic piece of art that he has created. His goal in this has been not only precision, but acutely unique visuals, something that grabs the senses and doesn't let go until every last detail captures the wearer, a process that may take years. "The customer always buys so much more than he or she initially thought," he points out on how details make themselves known with time.

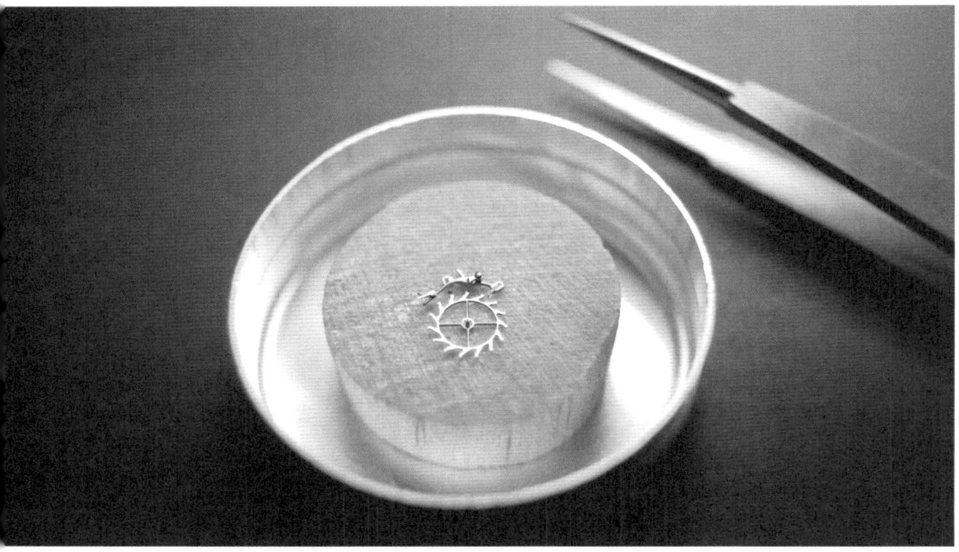

"Our watches are for people who are all about content, not brands," the master watchmaker, who grew up in the Thun area, says in his serious way–which is most likely why there is only a very thin line between what he does for a living and how he lives, the borders dissipating as he moves from room to room within Nussbühl, one private, the next public.

"Those who no longer need labels and marketing are our customers," he continues. "From far away, my watch should look like any other, inconspicuous. Then the client can enjoy it in peace. If it were too conspicuous, they would no longer want it, for it would then awaken covetous feelings. It is much more beautiful as a secret pleasure for the wearer. This is what we make: secret pleasures for connoisseurs."

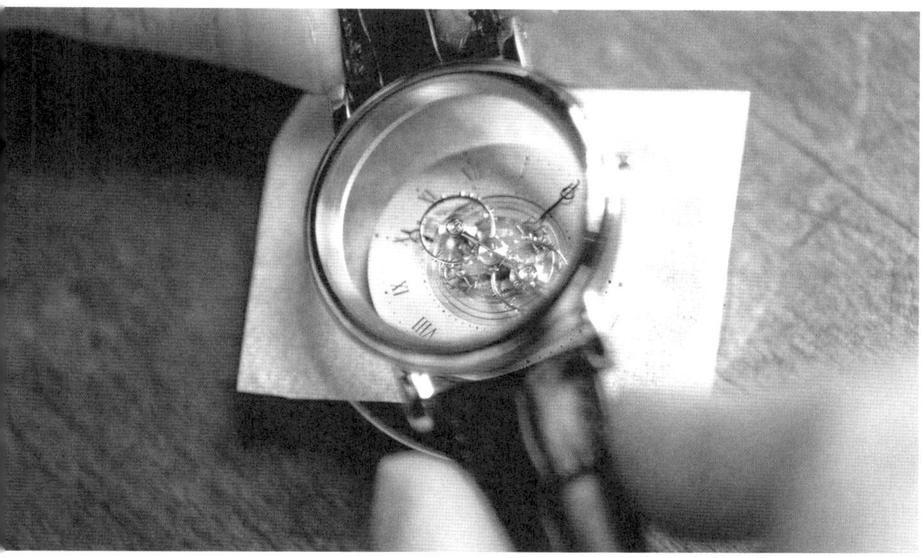

In front of Haldimann's stately home and place of business, which is endowed with a tremendous view of majestic Lake Thun and the Alps, the master has a set of sculptures that proclaim to the world what this man and his work are all about. These "fire sculptures" created by local artist Paul Widmer do what so many of Haldimann's watches and clocks do: they resonate. "This fits so well with our philosophy," he explains as he sets off the unique monoliths, which seem to play with each other, alternately tossing fire back and forth.

This is also what the clock seems to do that greets the visitor in the hall of the majestic house: two pendulums, crossing in the middle, swing in alternate routes–simultaneously driving two separate movements displaying local time and a second time zone with a 24 hour hand.

Having completed his first resonance clock in 2001, many of his movements now bend to the coupling of double escapements supporting each other, one making the other better and more accurate. While Haldimann–who only works with traditional techniques and does not even own a CNC machine–admits this duality could well cause them to destroy each other, he notes with a knowing smile that such is not the case. "I'm so interested in why it works," he says, the statement a tip of the iceberg in his study of the phenomenon of resonance.

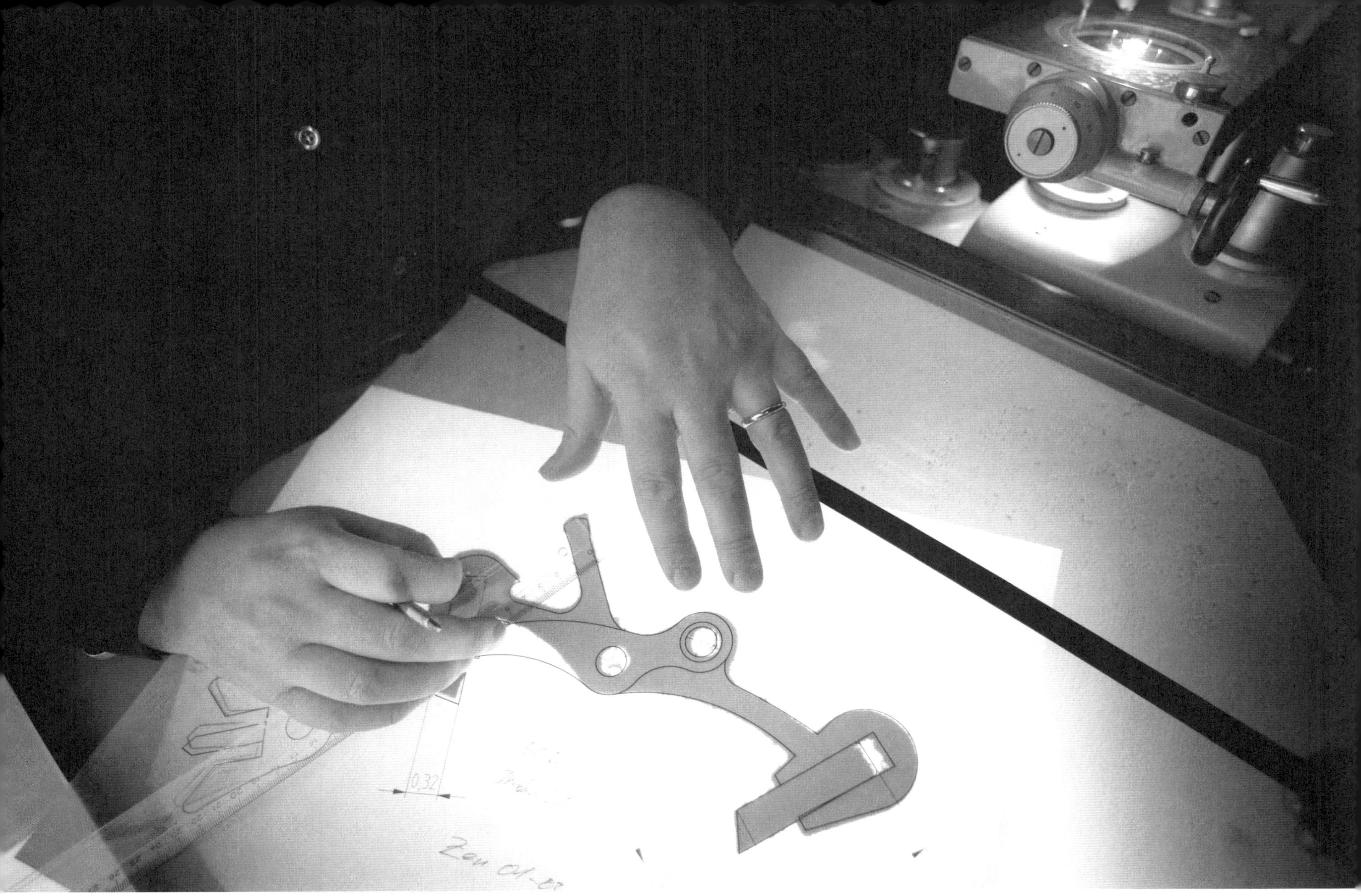

A flying tourbillon is what constitutes the resonance of his H1 wristwatch: while this movement contains only one escapement, Haldimann explains that “it is the cage and the tourbillon that resonate with each other and begin to sing.” This topic is endless for him, the tourbillon as “a sculpture in itself,” a melody that needs to be released.

Another of Haldimann’s interesting timepieces does not display the time at all: though the H8 came about at the request of a client, this wearable sculpture with no other purpose than to exist as portable art fits right in with Haldimann’s consistent way of being. “We will all finally get to the point where we no longer need the time,” he explains as the concept. “It is actually the greatest symbol of wealth when you no longer need to know. The essence of all my timepieces can be found in the H8, the Skulptura.”

This essence germinates here at Nussbühl, where time seems to stand still and the tick, tick, ticking of the clocks creates an audible backdrop to the nearly invisible accoutrements that are the base of Haldimann's melodic art form.

“Recognizable only under a microscope, what we do here is at the limit of what a person can create by hand.”

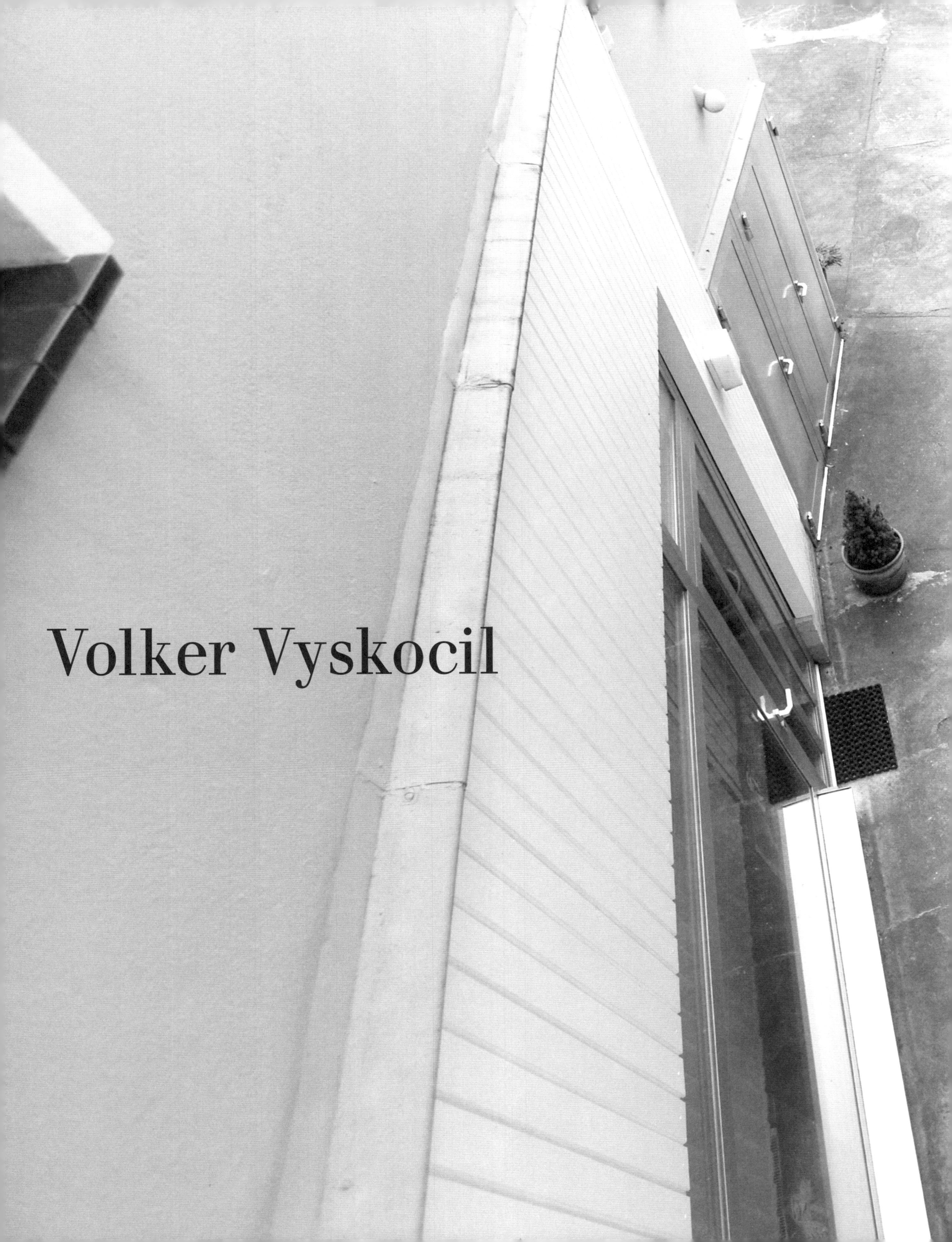

Volker Vyskocil

Unorthodox: this is the perfect word to describe every facet of Volker Vyskocil. Unlike practically every other watchmaker working at this level of complexity, Vyskocil did not learn watchmaking by the standard route. Classic methods don't apply to him, and school was not his thing, though he later developed the ambition to find his own way of attaining the technical university education that he desired. During this time in his life, he instead discovered he had a relationship with machines and even invented a new one–which unconventionally led to a career in mechanical engineering.

However, ascertaining that the 9-to-5 lifestyle was not one he wanted to have for keeps, he risked changing gears to make a living from an object that had fascinated him from childhood. Back in his teens, Vyskocil had originally wanted to become a watchmaker, but the timing was not right in that pre–mechanical renaissance era. Additionally, Vyskocil is at home in a region that does not necessarily have a tradition in the craft. Nettetal is situated ten kilometers from the Dutch border in Germany's Rhine region, and while the population is constituted from a good mix of these cultures, mechanical watchmaking has no foothold there.

Without a master watchmaker's title, and, indeed, without a watchmaker's education at all, Vyskocil taught himself all he needed to know about designing and making movements. The rest simply came naturally, an inherent extension of the combined elements that create his being.

Vyskocil is nothing if not sparing in his ways: he makes use of every millimeter, whether it is his personal living space, his workshop elements, or his movements. And what is even more interesting is how all three of these components blend to form a single unit. Vyskocil's apartment is actually also his workshop, with little machines scattered all over the place: polishing wheels are right next to the washing machine; galvanic baths can literally be found in the Vyskocil bathroom; a miniature kiln for firing enamel dials is located in a storage closet, and his CAD-outfitted computer is on a desk in the room that passes for an office, which he shares with his wife, who is highly involved in the organizational side of his business. Only one room is entirely dedicated to Vyskocil's horological workshop, and it is the only room that looks anything like a traditional watchmaker's atelier–this thanks only to the lathe, bench, profile projector, and other equipment necessary for the creation of wristwatches found there.

What catches the eye first in this busy little room is without a doubt the homemade CNC machine that Vyskocil designed and built himself to manufacture components like base plates. This unique tool is indicative of the unorthodox niche Vyskocil has carved out for himself; the ten or so watches he makes every year for customers, who treat them almost as if they were their own children, are graced with a production depth rarely found anywhere else–and each part is the work of the master himself.

"I make almost everything myself, something that is actually illogical," Vyskocil explains of his self-imposed solitude. "It perhaps started because it had to be that way; it was impossible to find suppliers and I might have a problem when [outsourced parts] didn't turn out the way I wanted because I had to compromise. So I thought I would just rather do it myself. Whether it's better that way, I don't know, but it's the way I do it. It's like a painter who wouldn't outsource part of his picture to a supplier saying, 'paint me a beautiful field.' I paint my fields myself. I have my own brushstroke, don't I?"

Unlike his independent colleagues, Vyskocil in his determined way knew immediately what his destination in the world of *haute horlogerie* was. He didn't waste years on creating movements, complications, or watches for established companies or restoring vintage masterpieces for auction houses. He went right to the meat of the matter and set out to create his own unique movement. "You have to make sure you are different from all the others," the introverted engineer explains of his base concept. "You have to deliver an argument for why they [the customers] should like you, or your watch. That is actually only possible if you create something unique, something they can't get anywhere else."

S10
B14
ST8
ST5
B17
B18

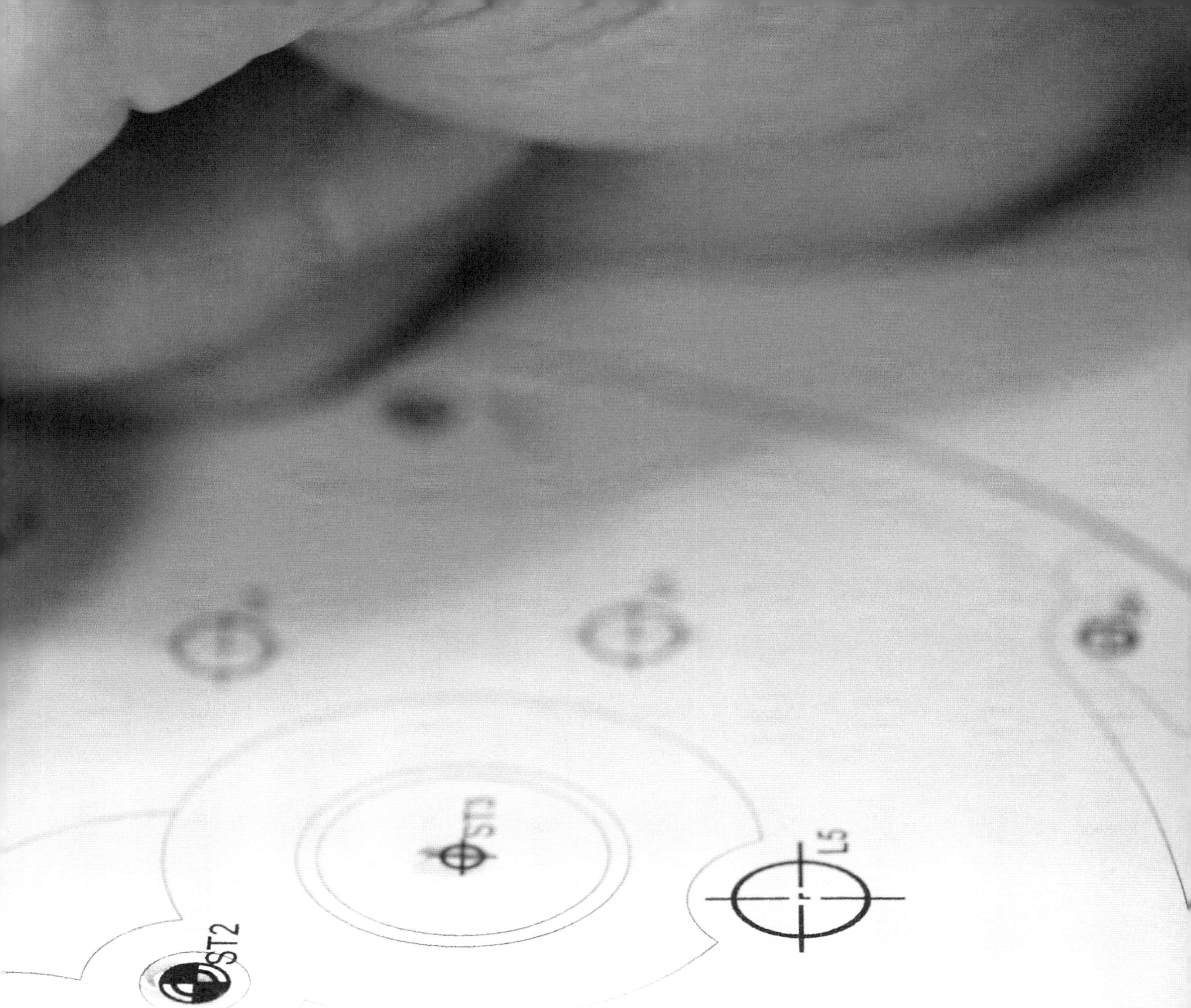

"It's like a painter who wouldn't outsource part of his picture to a supplier saying, 'paint me a beautiful field.' I paint my fields myself. I have my own brushstroke, don't I?"

Räum
&
Schneidöl
RS 331
Rotes
Blutlaugensalz
Wiener Kalk

FR. REINER
MÜNCHEN

Vyskocil doesn't hesitate to explain that he first looked to history for inspiration in beauty and precision, "Forty years ago people made beautiful watches. Then it became very industrial, and the main object became making reliable watches, but in high numbers. I like to think that forty, fifty years ago, the watchmaker had more to say."

Vyskocil, the personification of purity in mechanical technology, was faced with his own conundrum: he prefers large, heavy balance wheels, for example, an element that comes with a price. "How do I fit it in?" he asked himself. "I'll change the gear train and take the minute wheel out. By making so much room available, I can pump up the spring barrel and balance," he answered himself. Thus, he definitively created his own style according to the elements he found most significant. More important, his ideas began to come in rivers. "I no longer felt so desperately that I had to find something new; these then came automatically."

His novel ideas have thus far centered less on complications than on changes and innovations within the watch, a growth comparable to nature itself and completely in line with his own character. "And this is why I prefer to live in 'the provinces,'" he explains, "to keep the blinders on and not look at what the others are doing. I am not held hostage by tradition; I only try to adapt it where I think it fits or it expresses who I am."

Vyskocil is almost the archetype of the starving artist. "As soon as I begin to visualize myself having to make a large quantity of watches, it doesn't work for me. This is why I have made the conscious decision to work here alone, so I can really work like an artist," he confirms. A technical artist.

His path less traveled–which includes nonconformist, uncompromising visions–is tangible in the end result: classic steel, not modern stainless, is the material of choice. Finishing, yes, but not too much, decoration not being the focus, but rather the protection of rustable steel components; his sandblasted look projects a certain mechanical beauty and a style that is all Vyskocil. He even cuts threads by hand and makes the hinges and locks of the boxes that his watches are delivered in.

"I'm very emotional, both in perceiving and in expressing," says Thomas Prescher, who agrees that his feelings tend to run away with him at times, stretching from impulsivity to aggressive boldness and detailed care. This sensitivity is probably what allows Prescher to be so inspired by the work of past thinkers such as Abraham-Louis Breguet, Anthony G. Randall, and Richard Good–the likes of whom did indeed motivate him to achieve what no other has yet done: a trilogy of wristwatches containing three different types of tourbillon: single, double, and triple axis variations.

This emotionality also allows him to follow his visions, which he admits only someone ready to be absolutely committed to them should do. His visions have led him to not only establish his own independence, both in terms of his own company and a certain non-reliance on suppliers, but also to realize three wristwatches that only a handful of prodigious watchmakers have ever attempted to create.

Thomas Prescher

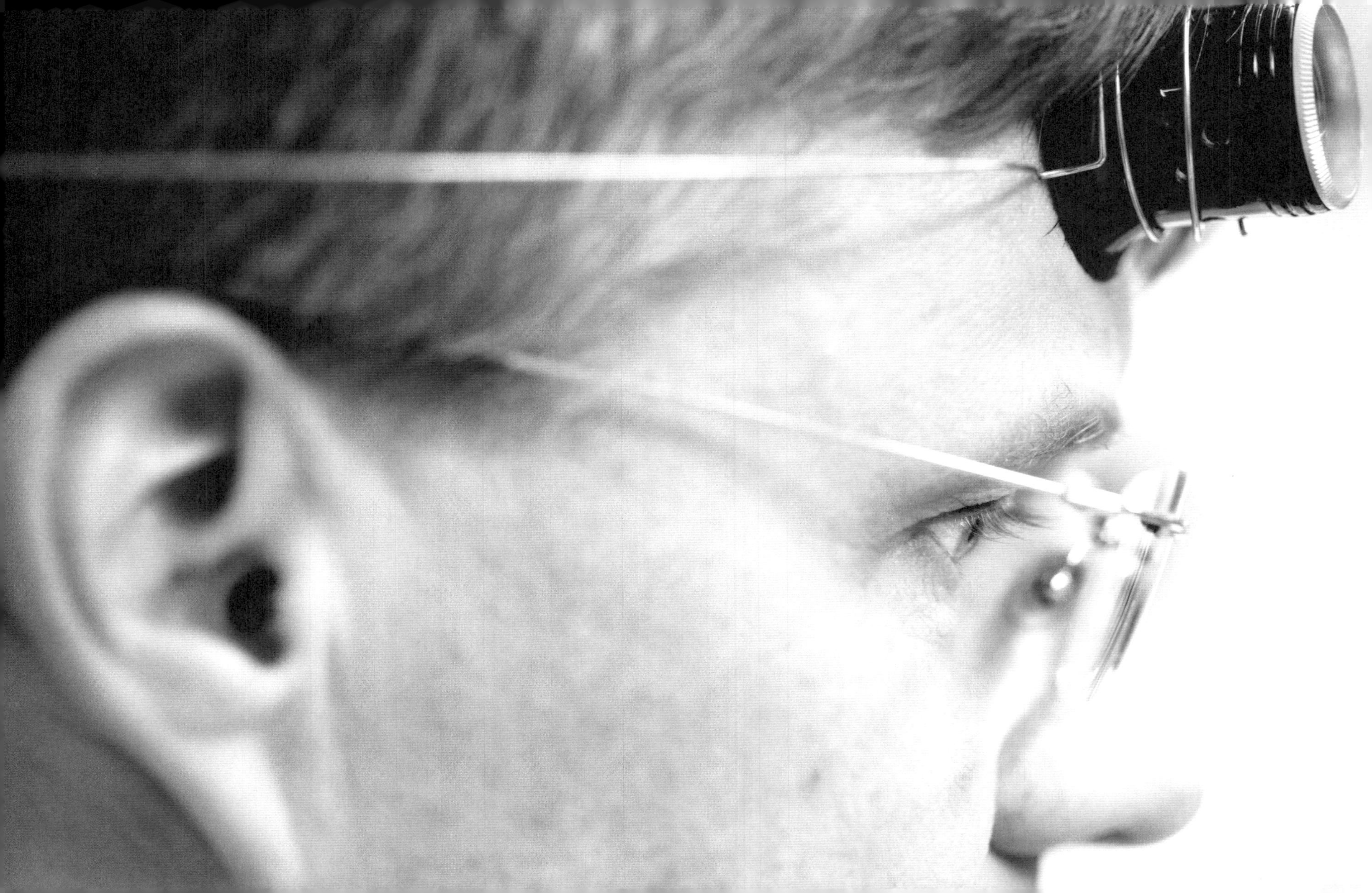

Prescher describes getting sudden inspirations that he feels he must follow. He is, however, very methodical in following them: “Ordnung muss sein” as the German saying goes, “everything has its order.”

In following his visions, he has not allowed his environment to negatively influence him. “Most great ideas get buried by outer influences,” the German-born master watchmaker explains, continuing with another German saying he is fond of, “geht nicht gibt’s nicht”: there is always a way. “We have always been able to find a solution, and my Triple Axis Tourbillon is the best example of this.”

His ability to follow his own sudden inspirations is well exemplified by the speed with which he actually completed a first working prototype of this complicated, transparent tourbillon. From the first line drawn on paper to this stage only six months had passed. “I create it in my head first. That’s where I see how it should be. It is already in the metal; you just have to free it.”

“It is already in the metal;
you just have to free it.”

During his career as a watchmaker, which he began only in his mid-twenties after his first spending six years as a navy captain, Prescher has gone through various horological stations, one of which led him down a restoration path similar to that which many of his independent colleagues have followed.

After a stint as production manager for a luxury watch manufacturer, he finally realized his need to do things his own way in order to realize his true calling. Prescher's own laws will not be disobeyed, however, so the freedom he now enjoys comes with its own set of mature rules, which he follows to a tee. "Creativity needs freedom, and it follows its own laws," he says. "Completion needs pressure."

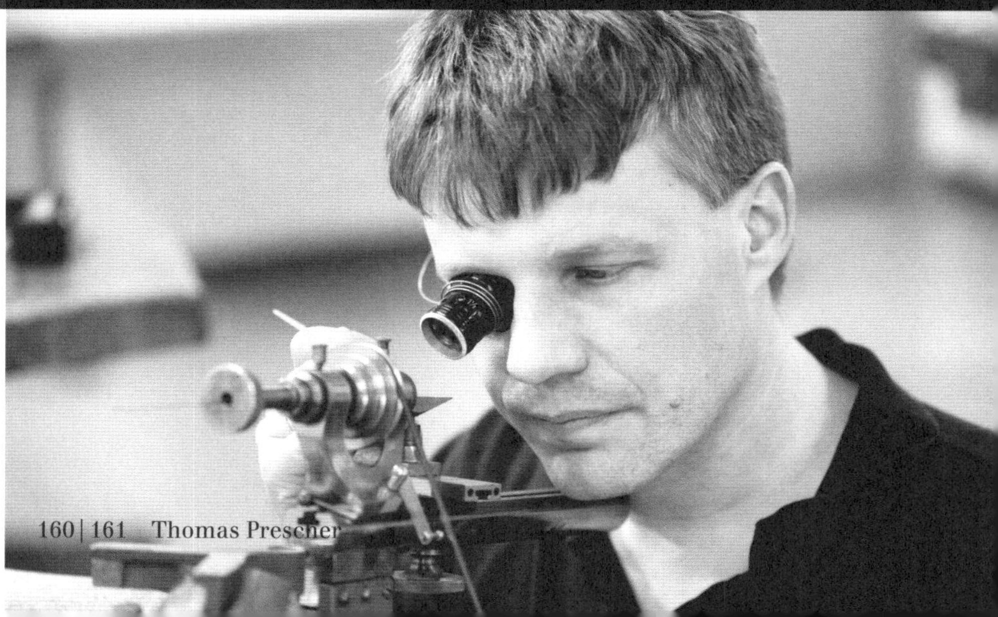

This might sound vaguely contradictory, but a sort of yin-yang is strangely prevalent within Prescher; there is no *this* without a certain *that*. Technology does not exist without art, light does not exist without shade, and freedom does not exist without limitations. Perhaps this is one reason Prescher is never quite satisfied with his masterpieces, continuing to work away at them long after another might have considered them done.

Never seeming to be finished with his work, he continues to mull over processes in his mind even after they are completed. One example embodies this mind-set: "Say I have thirty screws to polish. While I am doing that, I am thinking about how I can make them better. So if I get to number 28 and I figure it out, I have to start all over."

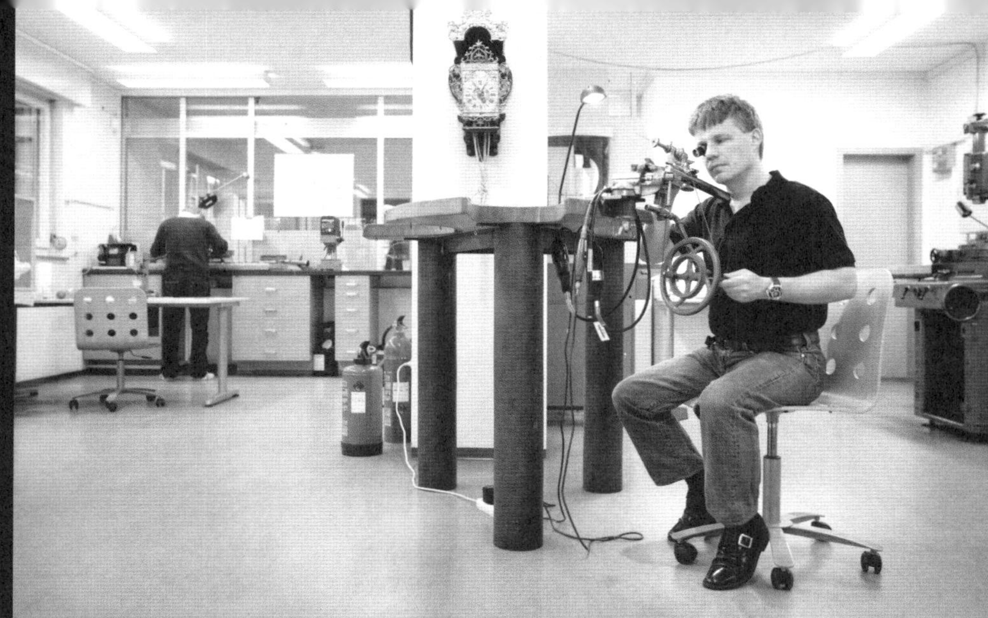

Some may call this an extreme case of perfectionism. Prescher thinks differently and tells a story about a key experience he had while still at watchmaking school. His assignment was to turn a stem, which he did and which brought him the highest grade. When the teacher asked him whether he could do a better job at it, he answered that he clearly could since both could see that the surface wasn't 100 percent satisfactory. Upon hearing this, the teacher broke the stem in two and told him to start over.

This ideal of excellence combined with a strict childhood and those six years in the military have certainly made him what he is today: efficient, goal-oriented, and demanding–both of himself and his team.

And while his goal is not, in fact, perfection—he does not hesitate to relate the fact that Muslim weavers purposely weave mistakes into rugs because to them perfection on earth does not exist and trying to achieve it is considered arrogant in the eyes of God—it is to create something that may well be termed kinetic art: beautiful, technical, and original. "Perfection doesn't exist," he states, though he does admit that it is his unique qualities of constantly searching and refining that make his pieces so interesting.

Getting his ideas to work motivates him as does making his visions and conceptions reality. "I ask myself 'why not?' instead of 'why?'" he says. "The best example of this is my Triple Axis Tourbillon. You do it and you know it's good. The world can't stop you from creating."

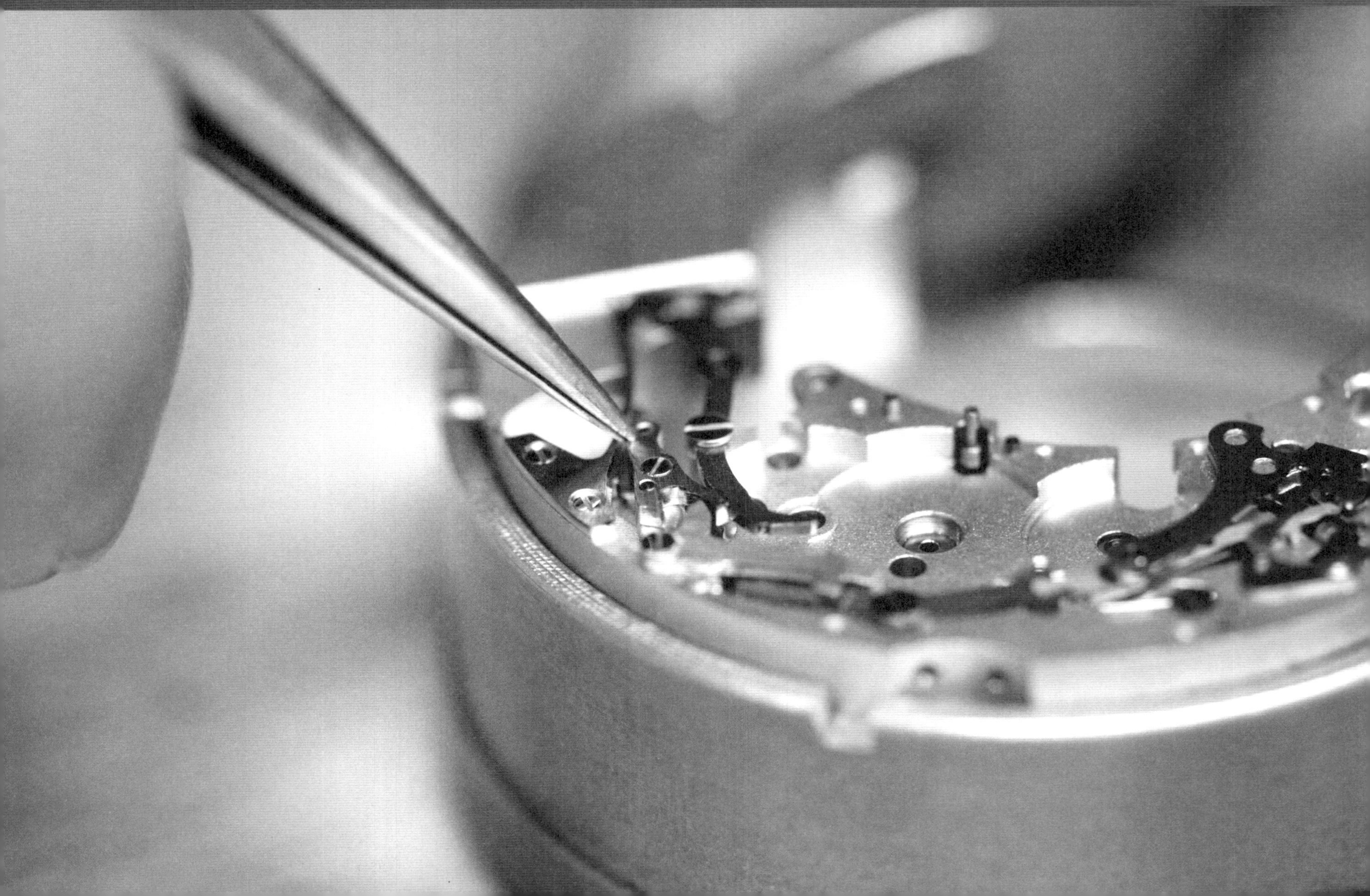

THOMAS PRESCHER

Prescher likes to recount how working with his hands and taking care of every part of a watch bring out the best in him, even though these are time-consuming processes. Driven and motivated, his joy is in creating a kinetic sculpture of mesmerizing beauty. “I don’t do it to finish it and to have created art. I love doing it.”

ROGER W. SMITH LTD

Roger Smith

An eighteen-year-old saves up and buys himself a quality watch, one known for reliability and robustness with a reputation transcending time and space. This is a big expenditure for a boy just finding out he's a man. Still a year away from graduating from watchmaking school, his future is uncertain, especially in light of the fact that the mechanical world has just emerged from the quartz crisis.

Perhaps this is why the boy is inspired to make a pocket watch himself. More likely the inspiration comes from George Daniels's visit to the Manchester School of Horology, where the young Roger Smith would receive his watchmaking diploma a year later.

"I couldn't believe a man could make a watch by hand," Smith remembers today. "From that moment on, I knew I wanted to make watches." His first pocket watch was more than just a timepiece, it was his proving ground: showing it to Daniels upon completion, he was gruffly told to go home and try again.

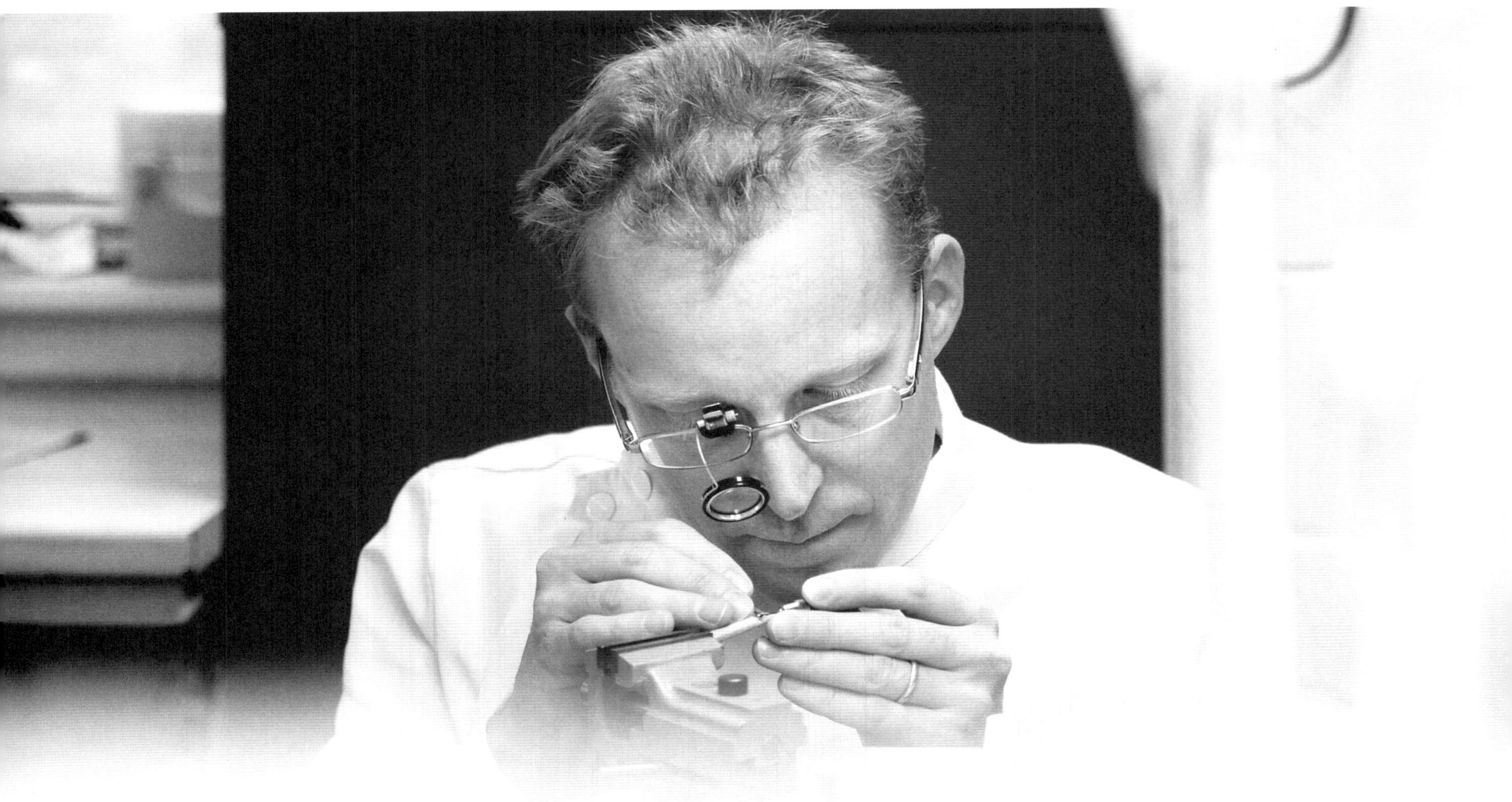

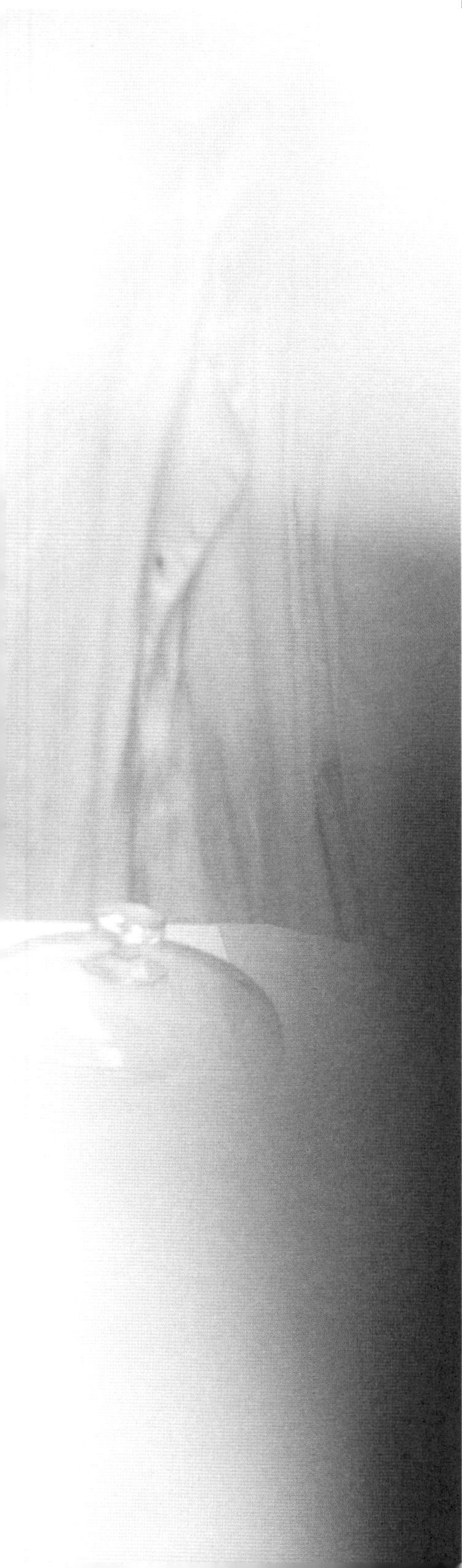

Though the ambitious pocket watch he had created even contained a working tourbillon, he followed this advice, showing up on the master's doorstep almost eight years later with a new version of his one-minute tourbillon embedded within a pocket watch he had completed alongside his day job in a service center of a popular watch brand.

The new work impressed Daniels so much that Smith became the only young watchmaker invited to share the master's workshop on the Isle of Man for any length of time, learning his art of elegant and masterful handcraftsmanship. Smith's humble goal was not to become someone future followers of horology might call "one of the masters," poignantly illustrated by his own lack of a master watchmaker title and almost reclusive desire against attracting attention. His aim was much simpler: "I wanted to make a watch that was up to the same sort of standard as the old pocket watches," he says today.

These days, Smith only dwells on the past when asked to do so or when contemplating how past master John Arnold may have treated a certain element. Having thoroughly explored the innovations and styles of the great pioneers of precision timekeeping in England–including those of mentor Daniels–he now makes watches his own way. The result is handmade timepieces that include a number of typical British elements, naturally, centered around the co-axial escapement that Daniels patented in 1980. “It can’t get more English than that,” Smith knowingly smiles.

“The long-term reliability will be greatly improved,” the young watchmaker explains as his other reason for wanting to use the rare co-axial escapement, ensuring his watches will stand the test of time. “And, yes, it is a more efficient escapement; the watches will keep a better rate over a longer period. And, from day one, it was very important to put style into the mechanism.”

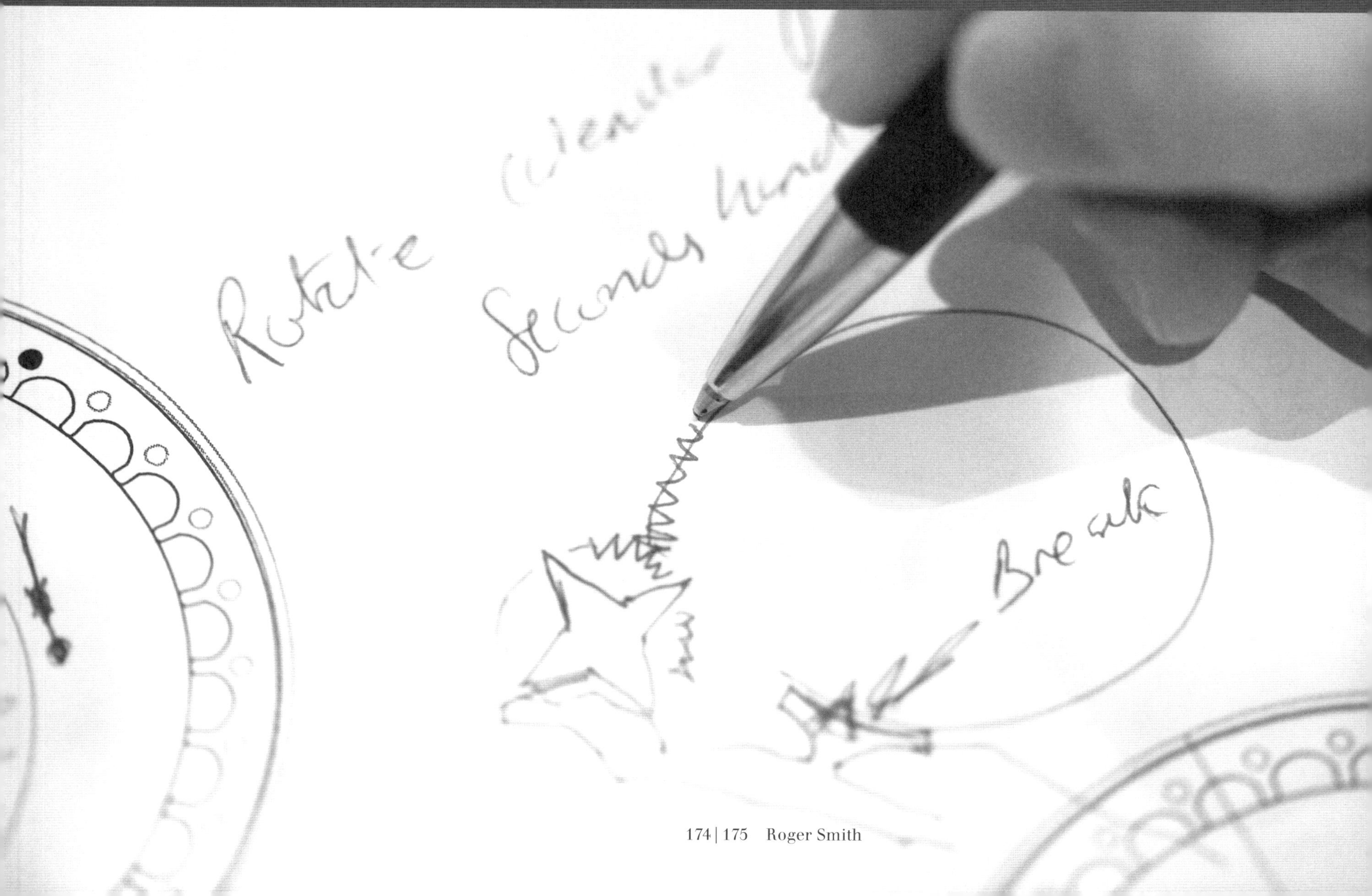

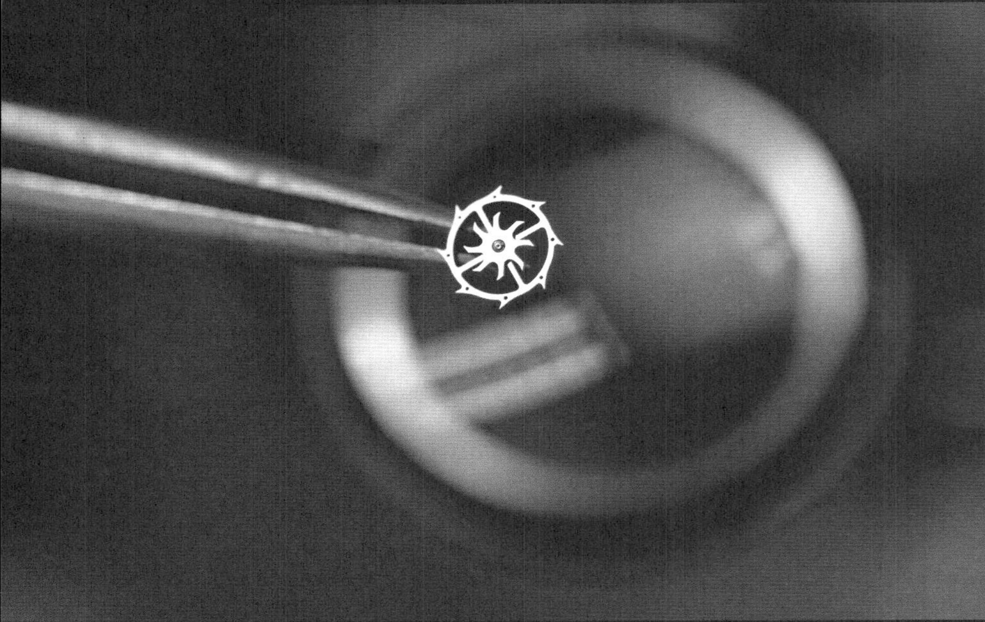

ISLE OF MAN
Mar-2008

Continuing the superb tradition of accuracy and precision set out by icons of British watchmaking such as Thomas Mudge and John Harrison–and continued in the modern era by George Daniels–Smith lives and breathes the English ideal of horology and a passionate desire to continue it. “History is untouchable: the British were pioneers of precision timekeeping,” he states simply and without questioning the whys and wherefores of his own focused interpretation of English timekeeping as it now descends through the ages.

In his work Smith does not limit his focus to the mechanism, finding the harmonious homogeneity of the entire piece to be more important than any single element. Therefore, the case, dial, hands, movement, buckle, strap, and even the box adhere to the concept he has set out for himself. "To me, every part is equally as important in wanting to create a great watch."

The tourbillon of his early endeavors is not the focus of Smith's world, though he will create one if requested by a client. Although these are seen by him and other horological experts as the highest expression in watchmaking, Smith prefers to keep his mind on the handmade character of his work. Therefore, his current baby, Series 2, is pleasurably pure and simple on the outside–while a closer look reveals the meticulous care that has gone into professionally manufacturing and finishing each and every element.

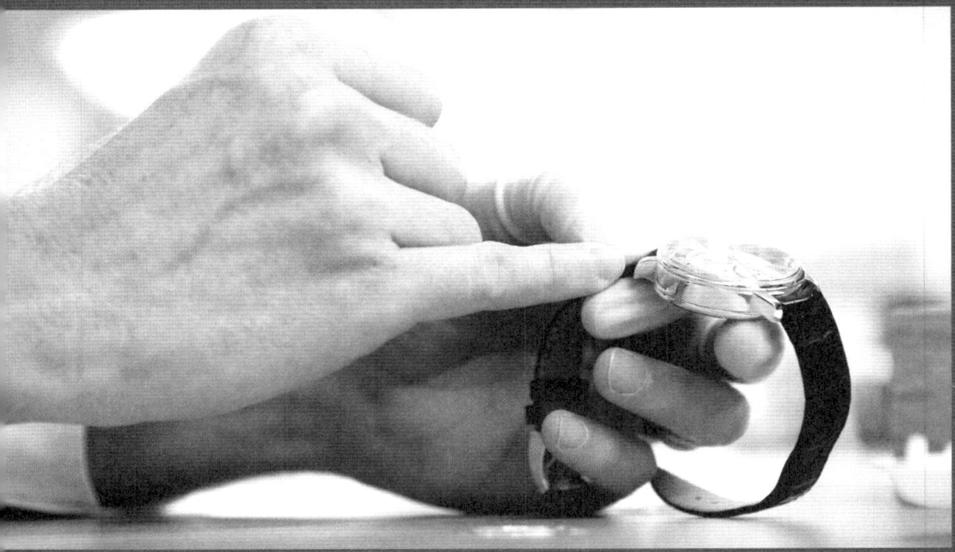

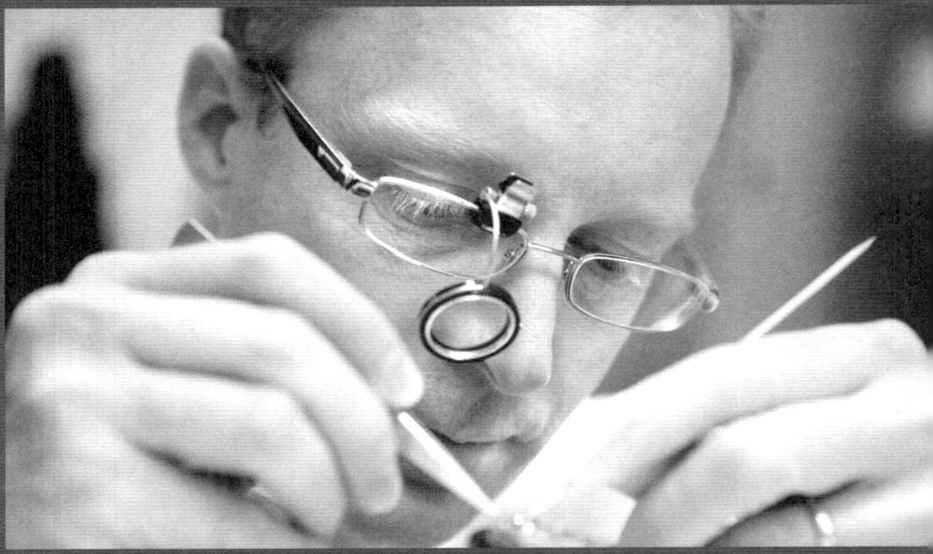

This is a world where beveling one lever can take up to half a day and the use of beautiful gold chatons on a frosted, gold-plated German silver plate serves both the aesthetics of the movement and its reliability. "It's all about trying to go to the highest level while keeping a hand-finished touch," Smith explains regarding his need for details such as three-dimensional hands inspired by old English pocket watches–the kind with real depth. "It's a personal approach. [My wristwatches] almost look like miniature pocket watches."

"I'm making my watches using materials that have been around hundreds of years."

"Traa dy liooar"–"time enough"–as the Manx would say. The inhabitants of Smith's chosen island home are used to the lush British paradise with its myriad accents resulting from an influx of various cultures and many shades of verdant flora interspersed with lonely rock.

This haven besotted with legend and myth may not be a traditional spot for watchmaking, but it certainly possesses many of the same characteristics today's more established horological locations have: peace, solitude, and a quietude that arises from absorbing the all-encompassing force of its natural environs. Perhaps it is even the windblown wildness of this island that inspires Smith to maintain his strict, disciplined style of technical artistry.

Felix Baumgartner

A third-generation watchmaker, Felix Baumgartner grew up with the sound of ticking, clicking, and whirring: his father maintained a houseful of clocks that proudly and continuously announced the time stored up inside of them. "It always went tick-tock everywhere at our house," he remembers of his not-so-distant childhood. "My father restores grandfather clocks, and that can be pretty arduous since at a quarter to twelve, the most imprecise of them would begin to ring midday. For half an hour, it always went boom, boom, boom. I grew up with this sound in my head."

Though he learned from the masterful timepieces that were the constant companions of his youth–“everything comes from one river,” Baumgartner doesn’t hesitate to point out–he has a distinct eye toward the future regarding his own creations. “We go so far with technology as to be dependent upon it: CNC technology, Teflon, titanium, various materials. Fifty, sixty years ago–without the precision of CNC machines–our watches simply would not have been possible. Urwerk is really an ensemble of machines and men. Most companies won’t say this so openly, but we are without inhibition; we want this,” Baumgartner describes as his take on the modern world of technology within the small company he has created with professional designer Martin Frei.

His is a generation that moves about in the world of technology easily and freely; a generation that never had to learn DOS or how to use a cell phone. Correspondingly, this is how Baumgartner approaches watchmaking: by taking advantage of every technical benefit he can; the more forward the technology, the better suited it probably is to his ultramodern timepieces.

URWERK

45
50 55 60
9
10 11 12

Audacious as they are, the mechanics that coin the look of Urwerk's timepieces surprise and amaze the observer at first glance: how can one stop looking for the new details to be found within the satellite system that lends the first collection of Urwerk's watches their incredible appearance? Baumgartner explains that this "newfangled" type of time display was literally based on history. "It was nothing so genius as waking up one morning and the idea was there," he self-assuredly explains. "In 1652, two Italian watch-makers presented something like this to the Pope, who wanted a clock he could read at night. What we did was to totally close the case, almost like an oyster, only allowing the time display to be seen. We have reinterpreted some-thing from the past, but added our vision and inspiration to it."

This bold new way of presenting traditional timekeeping has set off a trend: many new brands have popped into the world of watchmaking since Urwerk appeared on the scene thirteen years ago and proved that an audience for unusual horology exists. With the extreme confidence born of talent, Baumgartner explains that this is in fact no trend: this is just the tip of the iceberg of youthful enthusiasm, secure in the fact that so much more technology is on the way–tech-nology Baumgartner feels is destined to change the face of watchmaking.

"What we had envisioned in the mid-1990s is what is in our watches. At the time, the complicated watch world found that the more hands a watch had, the more complicated it looked, and the better it was. We did precisely the opposite: we had no hands." And with this "simple" concept, Baumgartner and Frei created their own culture within the high watchmaking scene, which at that time was not nearly as populated as it is today.

It wasn't always as easy as it is now, he continues, "In the 1990s it was like we were from Mars; people looked at us and kept going. They didn't understand. And all of a sudden, three, four years ago, it became accepted."

This is perhaps because a more sophisticated watch-buying public began to see where this young watchmaker and his even younger company come from; for Baumgartner this is a logical conclusion. "There aren't as many inventions as people always think. From various already existing elements, we bring certain things together and a new idea is born. Is that then an invention, or is it inspiration? Or is it perhaps a sophisticated combination of the two? At Urwerk, I would say it is the latter, reinterpretation using modern technology."

Philosophy is one thing, but what about the elementary mechanical prerequisite that needs to be intrinsic to a high-quality timepiece? "Our watches have more components than a tourbillon and could possibly be even more complicated than one," Baumgartner explains of his need for technology. "It is simply a complication for today, and therefore requires extreme precision. And new influences need to come into it, which is possibly one of the reasons Urwerk was able to develop so positively over the years."

Baumgartner, 33 years of age, is really at the beginning of his career even though he could already be considered an authentic fixture in the modern world of complicated watchmaking. Whether it is guts, experience, talent, or his anti-authoritarian nature, his artistic style is provocative, and intended to be so. "It is simply a reality that traditional watchmaking with traditional complications such as the minute repeater, the tourbillon, and the perpetual calendar have lost their logic and rationale. They have suddenly become *objets d'art.* A tourbillon is nothing more than mechanical art–especially since it is not more precise [than other forms of mechanical watchmaking]. This is only the beginning of this type watchmaking, nothing more than a petty beginning."

Over and above the pure mechanics, Baumgartner is sure that modern aesthetics and technology will continue to evolve in high watchmaking. "So much is still possible; we are simply the ones who had the guts to bring something new to the industry, a new aesthetic." He prophesies that even bigger companies are yet to be established that will continue in an avant-garde style, both inside and outside the watch. When you look at Urwerk, it's hard to imagine the future will bring anything more postmodern. Then again, possibilities are endless in Baumgartner's own youthful mind. "Electronics and new technologies will come into play and become history, and there will be fans of this technology as there are currently fans of tourbillons or Breguet hairsprings, or whatever. This is the natural process of history."

"This is really a product of today, but one with its roots in the history of the mechanical watch. There isn't much that is a completely new invention."

MOON

Philippe Dufour

Born
June 3, 1948,
in Le Sentier, Switzerland

Education
Watchmaker repairer *(rhabilleur)* diploma from the École Professionelle de la Vallee de Joux, 1967, today this school is known as the École Technique de la Vallée de Joux

Awards
Prix Gaïa 1998

Watchography
Audemars Piguet Grande et Petite Sonnerie pocket watch 1986,
Grande et Petite Sonnerie 1992,
Duality 1996,
Simplicity 2000

Paul Gerber

Born
January 31, 1950,
in Bern, Switzerland

Education
Watchmaker repairer *(rhabilleur)* diploma from the Gewerbeschule, Zürich 1970

Awards
Entry in the *Guinness Book of World Records* 1989 and 2006,
Prix Gaïa 2007

Watchography
Caliber 17, Piguet/Muller/Gerber creation 1992–2003,
Retrograd 1996,
Retro Twin 2001,
Model 33 2004,
MIH Watch 2005,
Ochs und Junior Settimana Junior 2009,
Ochs und Junior Anno 50 2009

Dr. Ludwig Oechslin

Born
February 10, 1952,
in Gabicce Mare, Italy

Education
Diploma in archaeology, ancient Greek, and Greek, University of Basel, 1976;
PhD in philosophy, scientific history, (theoretical physics), and astronomy, University of Bern, 1983; post-doctorate professorial qualification with *venia legendi* in technical archaeology, ETH Zurich, 1995,
Watchmaker repairer *(rhabilleur)*, 1984;
Master watchmaker, 1993

Awards
Prix Gaïa 1995,
Recognition Award of the City of Lucerne 1998,
Gold Medal of the Swiss Society of Chronometry 2002

Watchography
Ulysse Nardin's Trilogy of Time:
Astrolabium Galileo Galilei 1985,
Planetarium Copernicus 1988,
Tellurium Johannes Kepler 1992,
Ulysse Nardin GMT +/- Dual Time 1994,
Ulysse Nardin Perpetual Ludwig 1996,
Ulysse Nardin GMT +/- Perpetual 1999,
Ulysse Nardin Freak 2001,
Ulysse Nardin Sonata 2003,
MIH Watch 2005,
Ochs und Junior Settimana Junior 2009,
Ochs und Junior Anno 50 2009,
Ulysse Nardin Moonstruck 2009

François-Paul Journe

Born
March 22, 1957,
in Marseille, France

Education
watchmaker repairer *(rhabilleur)* diploma from horological school in Marseille, 1975, master certificate from the École d'Horologerie de Paris, 1976

Awards
Prix de la Fondation de la Vocation Bleustein-Blanchet 1987; Prix Gaïa 1994; Grand Prix d'Horlogerie de Genève, Special Jury Prize for Octa Calendrier 2002; Grand Prix d'Horlogerie de Genève Best Men's Watch, Octa Lune 2003; Grand Prix d'Horlogerie de Genève Aiguille d'Or, Tourbillon Souverain 2004; Grand Prix d'Horlogerie de Genève Best Men's Watch, Chronomètre Souverain 2005; Grand Prix d'Horlogerie de Genève Aiguille d'Or, Sonnerie Souveraine, 2006; Chevalier des Arts et des Lettres awarded by the French Ministry of Culture 2006; Grand Prix d'Horlogerie de Genève Aiguille d'Or, Centigraphe 2008

Watchography
Tourbillon pocket watch 1978,
Tourbillon pocket watch with constant force mechanism 1982,
Automatic chronometer pocket watch 1986,
Planetary pocket watch 1987,
Tourbillon wristwatch with constant force mechanism 1991,
Tourbillon Souverain 1999,
Chronomètre à Résonance 2000,
Harry Winston's Opus One 2001,
Octa Automatic 2001,
Octa Chronograph 2002,
Octa Reserve de Marche 2002,
Octa Calendrier 2003,
Octa Lune 2003,
Octa Zodiaque 2004,
Tourbillon Souverain dead seconds 2004,
Chronomètre Souverain 2005,
Vagabondage I 2005
Sonnerie Souveraine 2006,
Centigraphe Souverain 2007,
Répétition Minute Souveraine 2008
Chronomètre Bleu 2009
Vagabondage II 2009

Kenji Shiohara

Born
January 8, 1958,
in Shioriji City, Nagano Prefecture, Japan

Education
National Japanese qualification
First Class Skilled Artisan, 1995

Awards
Gold medalist World Skills Competition 1979, Medal of Yellow Ribbon from Japanese Emperor 2005

Watchography
Credor Spring Drive Skeleton 2003,
Credor Moon Phase with Spring Drive 2005,
Credor Sonnerie 2006,
Credor Eichi 2008

Kari Voutilainen

Born
June 2, 1962,
in Rovaniemi, Finland

Education
Watchmaker repairer *(rhabilleur)* diploma from the Finnish School of Watchmaking in Tapiola, Finland 1986;
Diploma from the Finnish Institute of Business Management 1988,
WOSTEP-Neuchâtel Refresher Course 1988,
WOSTEP-Neuchâtel Course on Complicated Watches 1989

Awards
Watchmaker of the Year presented by the Finnish Association of Watchmakers 2006, Grand Prix d'Horlogerie de Genève Best Men's Watch 2007

Watchography
Tourbillon pocket watch with power reserve 1994,
Tourbillon chronometer with constant force escapement 2000,
Decimal Repeater Masterpiece 6 2005,
Decimal Repeater Masterpiece 8 2005,
Decimal Repeater Masterpiece 7 2006,
Masterpiece Chronograph 2006,
Decimal Repeater Masterpiece 9
Tantalor 2007,
Observatoire 2007,
Chronomètre 27 2008

Vianney Halter

Born
March 24, 1963,
in Suresnes, France

Education
Watchmaker, *horloger complet*, diploma from the Ecole d'Horlogerie de Paris, 1981, *certificat d'aptitude professionelle en horlogerie*

Awards
Grand Prix d'Horlogerie de Genève Technical Innovation and Complication Award with Harry Winston for Opus 3, 2003

Watchography
Antiqua Perpetual 1998,
Contemporaine Perpetual 1999,
Classic 2000,
Contemporaine Moonphase 2001,
Goldpfeil Faces of Time
and unique piece 2001,
Trio 2001,
Harry Winston's Opus 3 2003,
Trio Grande Date 2006,
Cabestan Prototype, with Jean-François Ruchonnet 2006,
Classic Janvier Moon & Sun 2007

Beat Haldimann

Born
May 5, 1964,
in Bowil, Switzerland

Education
Certified clock and watch repairer *(rhabilleur)* diploma from the Grenchen/Solothurn School of Watchmaking, 1985,
Certified master horologist, 1993

Awards
Prix Gaïa 2009

Watchography
H1 Flying Central Tourbillon 2002,
H2 Flying Resonance 2003,
H8 2008

Volker Vyskocil

Born
September 26, 1964,
in Lobberich, Germany

Education
Diploma in toolmaking 1983,
Diploma in mechanical engineering from University of Duisburg, 1993,
Self-taught watchmaker

Watchography
VA, V-30/45-01-A 2004

Thomas Prescher

Born
March 14, 1966,
in Lingen/Ems, Germany

Education
Watchmaker repairer *(rhabilleur)* diploma from the Solothurn School of Watchmaking, 1994, Master certificate from the Karl-Geitz-Schule, Frankfurt am Main, Germany, 1997

Awards
Canton of Schaffhausen's Top Award for the Best Examination of the Year 1994, Association of German-Swiss Watch Manufacturers Top Award for the Best Examination of the Year 1994

Watchography
Single Axis Tourbillon pocket watch 1994,
Double Axis Tourbillon pocket watch 2003,
Tempusvivendi 2003,
Tourbillon Trilogy, set including single, double, and triple axis flying tourbillon wristwatches, 2004,
Triple Axis Tourbillon wristwatch 2004,
Sculptura Una 2006,
Triple Axis Tourbillon Regulator 2008,
Perpetual Calendar QP 1 2009

Roger Smith

Born
March 21, 1970,
in Manchester, England

Education
Watchmaker repairer *(rhabilleur)* diploma from the Manchester School of Horology, 1989

Awards
British Horological Institute Bronze Medal 1989

Watchography
No.01 tourbillon pocket watch 1992,
No.02 tourbillon pocket watch with perpetual calendar 1997,
No.03 pocket watch with 15-second remontoire 2001,
George Daniels Millennium wristwatch series 1998–2001,
Two tourbillon wristwatches with a simple calendar designed by George Daniels, signed Daniels London 2000–2004,
Series 1 2001–2003,
No.03 tourbillon with date 2003,
Series 2 2006

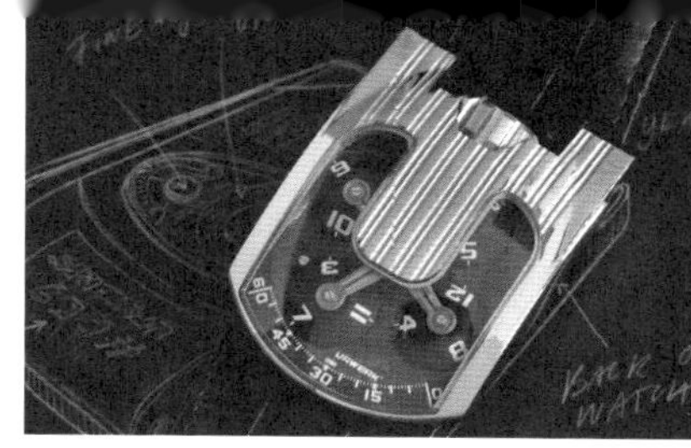

Felix Baumgartner

Born
August 8, 1975,
in Schaffhausen, Switzerland

Education
Watchmaker repairer *(rhabilleur)* diploma from the Grenchen/Solothurn School of Watchmaking, ZeitZentrum, 1995

Watchography
UR-101 1997,
UR-102 1998,
Goldpfeil Faces of Time and unique piece 2001,
UR-103 2003,
Harry Winston's Opus V 2003,
UR-103.03 2003,
UR-103 Blackbird 2006,
UR-103.05 Platinum 2006,
UR-103.06 Joaillerie 2006,
UR-103 special edition in collaboration with master engraver Jean-Vincent Huguenin 2007,
UR-201 2007,
UR-103.08 TiAlN 2008,
UR-202 2008,
UR-202 AlTiN 2008,
UR-103T Tarantula 2009,
UR-CC1 2009

Imprint

Text

Photography

Elizabeth Doerr
Karlsruhe, Germany

International watch journalist Elizabeth Doerr's expertise and standing in the field of watchmaking is unique: over the course of a twenty-year career, she has written for magazines all over the world, authored and coauthored books on watches, been a member of several international juries, and is the editor of *Wristwatch Annual*, a yearly overview of the mechanical watch industry.

Ralf Baumgarten
Cologne, Germany

Photographer Ralf Baumgarten has a varied repertoire of themes: his first book on the subject of watchmaking, *UhrMenschen*, received a Red Dot design award in 2005; in 2007 he spent time portraying a valley in Switzerland; and in 2008 he captured faces of camels in Dubai for a desert-themed project. His artful photography has been featured in publications and exhibitions in a number of countries–the latter including La Chaux-de-Fonds, Switzerland's International Museum of Horology.

Our thanks
go to the following people

The twelve watchmakers, without whose art we would never have been inspired, and their support teams supplying us with information including Brigitte Makhzani, Philippe Mariez, Yacine Sar, Kazuya Kawahara, Hiroaki Kobari, Shu Yoshino, Les Perry, Heike Prescher, Susanne Hurni, Ruth Gerber, and Heidi Vyskocil.

The very talented Gerd and Barbara Baumann, who patiently believed that art of this magnitude would triumph.

Ashley Benning, a most industrious, understanding, and professional copy editor. Thank you also to Tatjana Griffith, Henning Ohlsson, Claire Eder, Lydia Gauzer, Leonid Khankin, Rolf Duscha, Ian Skellern, Volker Schäffner, Thomas Hein, Markus Niesen, and Michael Fankhauser for various supportive aspects related to making this book a reality. Thank you to Hendrik teNeues, Karl Klose, Nele Jansen, and Nicole Sieben of teNeues Verlag GmbH & Co. KG for their continued interest in the watch industry.

And last, but never least, thanks to our respective supportive partners, Dr. Armin Doerr and Christel Mayer-Baumgarten, and of course our inspiring children Alexander & Sabrina Doerr and Lotte & Lene Baumgarten.

Design

Published by

Baumann & Baumann
Büro für Gestaltung
Schwäbisch Gmünd, Germany

This team is known for its work with the German parliament in Bonn, the Great Court of the British Museum in London, and internationally operating companies such as DaimlerChrysler, Siemens, and WMF.

Many other smaller initiatives and cultural or social facilities also bear the team's signature, like the corporate design of the Stetten Deaconry and Marbach's Schillerhöhe.

Barbara and Gerd Baumann often speak on their subject domestically and abroad and hold guest professorships, among them at the Musashino Art University in Tokyo.

The Baumanns have enjoyed a number of extensive publications in magazines and books and have been continuously awarded national and international prizes throughout their careers, among them the State Award of the Republic of Germany for corporate design.

www.baumannandbaumann.com

Foreword by Elizabeth Doerr
Text by Elizabeth Doerr
Design by Baumann und Baumann
Production by Nele Jansen,
teNeues Verlag
Color separation by
ORT-Medienverbund, Krefeld

Bibliographic information published by the Deutsche Nationalbibliothek.
The Deutsche Nationalbibliothek lists this publication in the Deutsche Nationalbibliografie; detailed bibliographic data are available in the Internet at http://dnb.d-nb.de.

Published by
teNeues Publishing Group

teNeues Verlag GmbH + Co. KG
Am Selder 37, 47906 Kempen, Germany
Phone: 0049-2152-916-0
Fax: 0049-2152-916-111
e-mail: books@teneues.de

Press department: Andrea Rehn
Phone: 0049-2152-916-202
e-mail: arehn@teneues.de

teNeues Publishing Company
16 West 22nd Street, New York, NY 10010,
USA
Phone: 001-212-627-9090
Fax: 001-212-627-9511

teNeues Publishing UK Ltd.
21 Marlowe Court, Lymer Avenue, London, SE191LP,
Great Britain
Phone: 0044-208-670-7522
Fax: 0044-208-670-7523

teNeues France S.A.R.L.
39, Rue, des Billets, 18250 Henrichemont,
France
Phone: 0033-2-4826-9348
Fax: 0033-1-7072-3482

www.teneues.com

ISBN 978-3-8327-9373-9

Printed in Italy